Arc Hydro Groundwater

GIS for Hydrogeology

Gil Strassberg

Norman L. Jones

David R. Maidment

Esri Press

REDLANDS, CALIFORNIA

Esri Press, 380 New York Street, Redlands, California 92373-8100

15 14 13 12 11   1 2 3 4 5 6 7 8 9 10

Printed in the United States of America

Library of Congress Cataloging-in-Publication Data
Strassberg, Gil, 1972–
Arc hydro groundwater : GIS for hydrogeology / Gil Strassberg, Norman L. Jones, David R. Maidment.
p. cm.
Includes index.
ISBN 978-1-58948-198-5 (pbk. : alk. paper)
1. Groundwater—Data processing. 2. Hydrogeology—Data processing. 3. Groundwater—Geographic information systems.
4. Hydrogeology—Geographic information systems. I. Jones, Norman L., 1961– II. Maidment, David R. III. Title.
GB1001.72.E45S77 2011
551.490285—dc22   2010037793

Ask for Esri Press titles at your local bookstore or order by calling 800-447-9778, or shop online at www.esri.com/esripress. Outside the United States, contact your local Esri distributor or shop online at www.eurospanbookstore.com/Esri.

Esri Press titles are distributed to the trade by the following:

In North America:

Ingram Publisher Services
Toll-free telephone: 800-648-3104
Toll-free fax: 800-838-1149
E-mail: customerservice@ingrampublisherservices.com

In the United Kingdom, Europe, Middle East and Africa, Asia, and Australia:

Eurospan Group
3 Henrietta Street
London WC2E 8LU
United Kingdom
Telephone: 44(0) 1767 604972
Fax: 44(0) 1767 601640
E-mail: eurospan@turpin-distribution.com

Table of Contents

Foreword

FEW GIS COMMUNITIES HAVE BEEN BRAVE AND WISE ENOUGH TO DEVELOP A SET of common procedures and approaches that can be widely shared and adopted. The hydrology community has been one of the most successful at accomplishing this. This is significant because the adoption of common approaches will become increasingly important as GIS moves onto the Web.

The work contributed by the hydrology community—closely tied to the science of water and water-resource management—is groundbreaking and revolutionary because it has led to a series of consistent, science-based methodologies for applying GIS to hydrologic applications anywhere.

Early on, leaders in hydrology realized that this work was essential for cataloging and understanding the quality and availability of freshwater that we use as humans and upon which all life on our planet relies. They began to think about developing a standard approach that could be used across a broad range of water-resource issues and geographies. They realized that to understand these precious resources, they needed to build rich content that was simple to use, standardized, and well-defined. It had to be information that was trusted and collected consistently across many organizations and jurisdictions—a continuous, harmonized GIS of the world's water resources.

Initially, their work and progress focused on surface water. A key breakthrough a decade ago was the development and release of Arc Hydro—a set of information models and tools for applying GIS to hydrology. These efforts, led by David Maidment, provided intellectual breakthroughs on how to apply GIS to water-resource problems. This led to the adoption and application of GIS by a large and growing scientific community. The results of their work have been applied across thousands of water organizations worldwide.

The Arc Hydro Groundwater data model presented in this book represents another major breakthrough in hydrology applications using GIS.

As the original Arc Hydro data model was being deployed, many members of the hydrology community knew that they wanted to integrate aspects of groundwater with their surface water modeling tools and solutions. Gil Strassberg, mentored by Maidment at the University of Texas, and Norm Jones and his team at Brigham Young University and Aquaveo, have made major advancements in their GIS database designs and methods for groundwater.

They initially focused on creating groundwater models and tools, but these were designed to work independently of surface water models and solutions. In their second stage, the authors began to articulate a vision in which these two worlds of hydrology (surface water and groundwater) could be integrated and combined into a unified view for water-resource management. After all, wasn't the ability to integrate a key GIS premise?

This book presents their results. It is the culmination of many years of hard work, and the result is fantastic. Their contributions include: the use of 3D information and visualization; time-series recording and modeling of observations; a design for operational information for groundwater management; the ability to integrate widely adopted, scientific groundwater models into GIS; and the use of GIS to support groundwater modeling applications.

Their designs are also elegantly simple and clearly articulated. Any hydrologist can understand and apply these models.

Meanwhile, this team continues to make progress in advancing GIS-based hydrology. And this dynamic and ever-changing process comes at a very exciting time. GIS is advancing into the new Web 2.0 world with all of its social networking implications. Three-dimensional GIS and dynamic, time-aware GIS are starting to be used in many fields. GIS and sensor information are being integrated into these environments as Web services that can be aggregated and combined using GIS.

This work continues to provide leadership on how to move forward on rich information models, analytical tools and results, interactive maps, 3D visualization, and working with time-based observations.

The hydrology community has assumed a leadership role for many exciting advancements in the application and use of GIS. Strassberg and his collaborators have captured these designs and methods. This book provides a foundation and a point of departure for future work and progress in this field.

I can't wait to see the progress they will make in the coming years.

Clint Brown
Esri
Redlands, California
October 2010

Preface

GEOGRAPHIC DATA MODELS ARE IMPORTANT BECAUSE THEY ESTABLISH A COMMON language enabling us to describe aspects of our environment systematically and consistently over large areas. Instead of having finite projects that end with specific outputs, the data model process for groundwater continues through time and across people and institutions to produce a comprehensive system that increases as we add information and detail. Water is ubiquitous in our environment and is vital to every human and life form on our planet. So many different factors bear on the management and science of water that the contributions of many disciplines are needed to describe and understand it. Hydrologic and hydrogeologic information systems built using standardized data models capture some of this variety of knowledge and information for everyone to share and build upon.

Arc Hydro: GIS for Water Resources, published by Esri Press in 2002, defined a geographic data model for surface water resources implemented in ArcGIS called Arc Hydro. In the years following, the Arc Hydro toolset has expanded to include more than 100 tools, becoming the most widely used geospatial toolkit for application in water resources. A question frequently asked during this period was, "What about groundwater?" This book responds to that question by describing a new and evolving geographic data model for groundwater named Arc Hydro Groundwater. As we presented the draft design for this data model to potential users, their feedback asked us to "connect surface and groundwater—they are really one resource." We have done this by creating a common framework data representation of surface and groundwater features, associated with a set of more specialized data model components that can be individually selected for particular aspects of groundwater resources. We hope that this process will inspire others to develop additional data model components to cover subjects not dealt with in our work.

The main goals of *Arc Hydro Groundwater* are:

- To provide a standard way to apply GIS to groundwater systems using readily available data such as geological and aquifer maps, locations of wells and associated tables of subsurface borehole information, and water observations data such as piezometric head levels and water quality measurements;

- To build a 3D representation of the subsurface hydrogeologic environment, including the vertical structure of boreholes, "geovolumes" representing the spatial extent of hydrogeologic units, and cross-sections drawn between boreholes or cut through the subsurface hydrostratigraphy;

- To take sparse groundwater observations in space and time and enable time-averaging and spatial interpolation of these measurements to create time-sequenced maps;

- To provide a link between ArcGIS and groundwater modeling, in particular to the MODFLOW (modular finite-difference flow model) groundwater simulation model that is a standard in the groundwater field.

It can thus be seen that *Arc Hydro Groundwater* overcomes two challenges normally limiting the application of GIS to groundwater systems—describing a 3D system in a 2D mapping environment and describing the time-varying properties of the groundwater system. *Arc Hydro Groundwater* resolves these issues by extending the 2D static GIS approach to a 3D description of subsurface hydrogeology and by providing space-time datasets for describing the time-varying properties of groundwater systems.

Arc Hydro Groundwater can be used to publish digital descriptions of groundwater systems using ArcGIS Server and to assemble the description of such systems using ArcGIS Desktop. As geospatial data becomes increasingly available for Web services, Arc Hydro Groundwater offers an ideal way for federal, state, and regional groundwater agencies to publish descriptions of their groundwater resources using a common structure.

We hope that this book, *Arc Hydro Groundwater: GIS for Hydrogeology*, will be valuable to readers from a variety of disciplines, including water-resource engineers, hydrologists, hydrogeologists, geographers, and GIS specialists. We believe educators will also find this book a valuable academic resource at the college and postgraduate levels. This book uses groundwater datasets from various locations in Texas to illustrate the application of *Arc Hydro Groundwater* to real situations. These datasets are available for download from the Arc Hydro Resource Center: `http://resources.arcgis.com/archydro`. Also accessible at this location are the graphics used in each chapter set up as PowerPoint presentations so that instructors using *Arc Hydro Groundwater* in their courses can readily incorporate visual material from the book into their lectures. A freely available toolset for getting started with *Arc Hydro Groundwater* is also available from the Arc Hydro Resource Center.

The major credit for designing and developing the groundwater data model goes to Gil Strassberg, who accomplished this as part of his doctoral studies at the Center for Research in Water Resources of the University of Texas at Austin under the supervision of coauthor David Maidment. Strassberg experimented with 3D data structures, examined what others had done in related fields, designed the data model components, built a prototype toolset to implement the data model, developed working examples of applications in various areas, and wrote his dissertation to describe all this research. Strassberg carried the major load as lead author in writing this book. The groundwater data model received Esri's Prize for Best Data Model at the 2006 Esri User Conference.

As work on the data model progressed, it became apparent that we needed to deal with two distinct bodies of knowledge—hydrogeology and groundwater modeling. Hydrogeology is a branch of geology dealing with water-bearing rock units near the earth's surface. Groundwater modeling is a branch of civil engineering that deals with management of groundwater resources and is concerned with issues of water supply and groundwater contamination. These are two very different perspectives, and we were fortunate in having advisers to guide us along the path—for hydrogeology, Randy Keller of the University of Oklahoma, and for groundwater modeling, coauthor Norman Jones of Brigham Young University. Jones, in particular, has been closely involved in the groundwater data model design and is a coauthor of this book. He has more than twenty years of experience in the groundwater field and is the principal developer of the widely used system that provides pre- and postprocessors for many groundwater models.

The design of the data model is important, but realizing all of its benefits requires an associated set of tools. Accordingly, Esri has partnered with Aquaveo LLC of Provo, Utah, to develop the Arc Hydro Groundwater tools under Jones' direction. The groundwater toolsets include a rich suite of tools for managing and analyzing groundwater data, including tools for querying, plotting, and mapping time series; editing 3D hydrogeologic models from borehole data, cross-sections, surfaces, and volumes; and for editing and visualizing MODFLOW simulation models. As a set of geoprocessing tools, the toolkit ensures that users will be able to build custom solutions and workflows. Portions of the toolkit are freely distributed, and portions are fee-based. These fees support ongoing tool development and management of the data model.

We thank Timothy Whiteaker and Steve Grisé, who participated in designing the groundwater data model and wrote and edited substantial parts of this book. We are grateful to reviewers Todd Jarvis, associate director, Institute for Water and Watersheds, Oregon State University; Eileen Poeter, director, International Ground-Water Modeling Center, Emeritus Professor of Geological Engineering, Colorado School of Mines; and Wei Li, research fellow, Center for Applied GeoScience, Hydrogeology, Eberhard Karls Universität Tübingen, Germany, for their valuable feedback on the manuscript. As always, we have been guided by our colleagues at Esri, and in particular we would like to acknowledge the contributions of Steve Kopp, Lori Armstrong, Michael Zeiler, Dean Djokic, Nawajish Noman, Zichuan Ye, Scott Morehouse, Clint Brown, and Jack Dangermond, who in various ways have supported and guided our work. Joe Breman, formerly of Esri, also was a valuable resource. And, to our friends at Esri Press, we acknowledge another debt of gratitude—Esri Press books are not just books, they are works of art in their own right. Thanks to Mark Henry, who edited the text, and to Peter Adams, Judy Hawkins, David Boyles, Monica McGregor, Brian Harris, and their colleagues for all their contributions to our manuscript and its finished product, this book.

Gil Strassberg
Norman L. Jones
David R. Maidment

Arc Hydro Groundwater Data Model

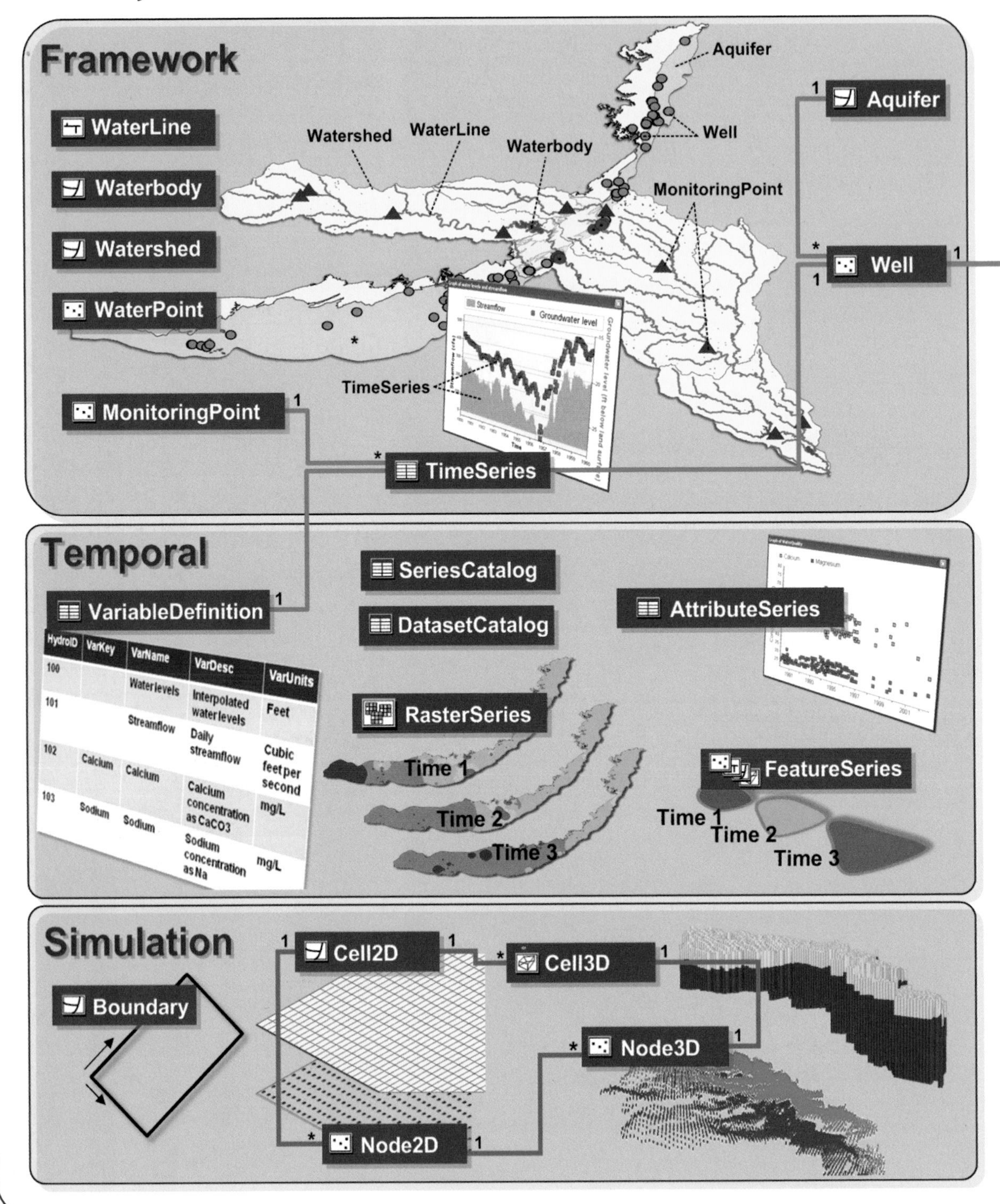

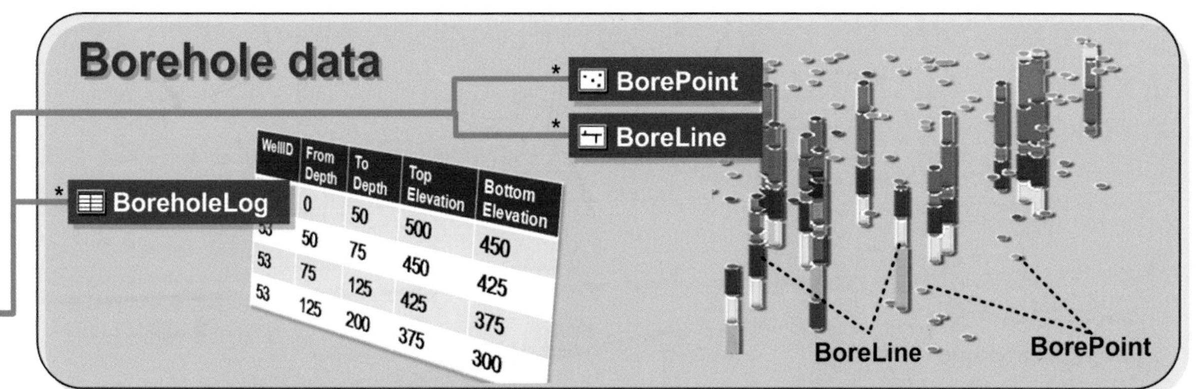
Borehole data
BorePoint
BoreLine
BoreholeLog
WellID
From Depth
To Depth
Top Elevation
Bottom Elevation
BoreLine
BorePoint

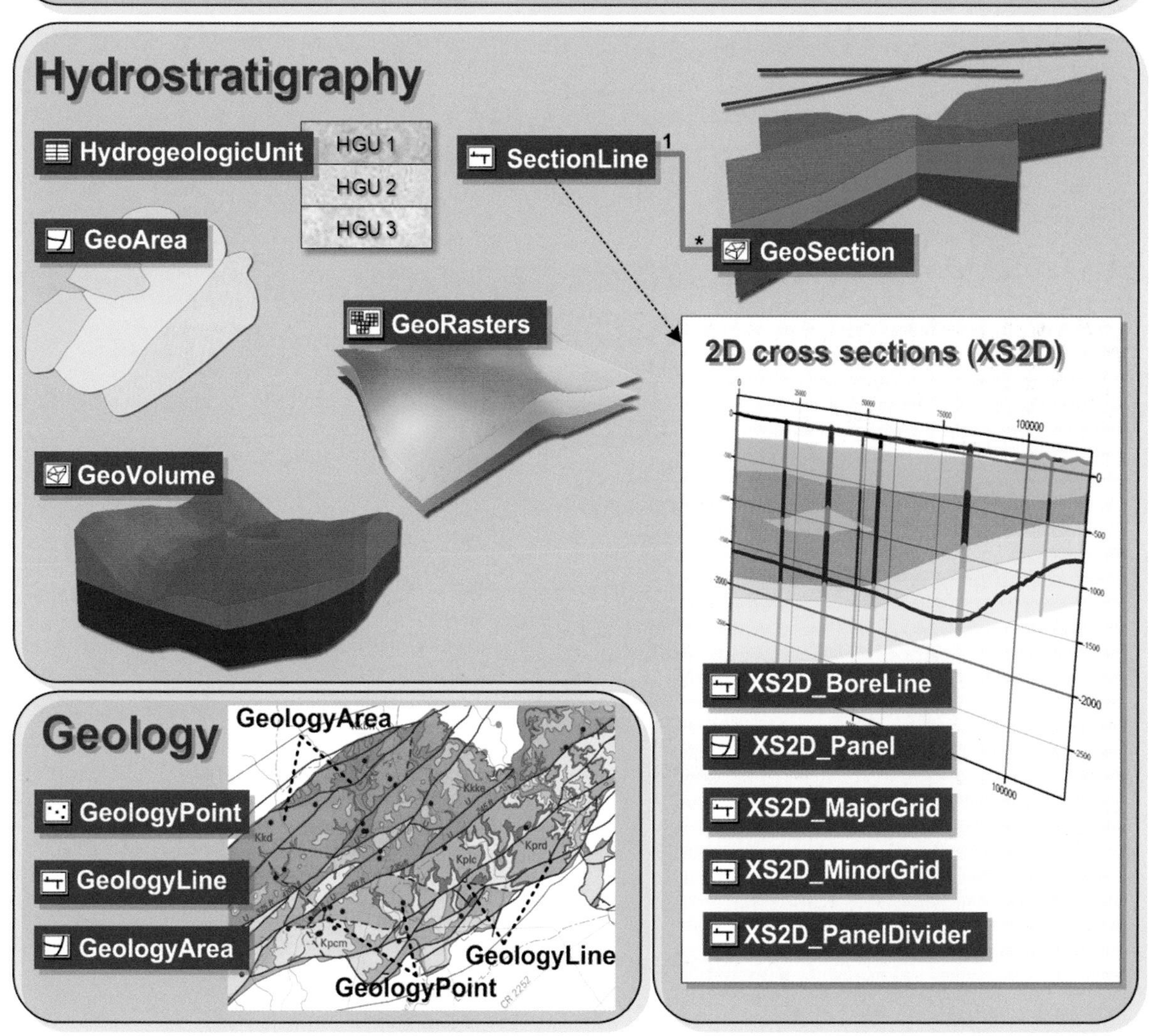
Hydrostratigraphy
HydrogeologicUnit
HGU 1
HGU 2
HGU 3
SectionLine
GeoSection
GeoArea
GeoRasters
GeoVolume
2D cross sections (XS2D)
XS2D_BoreLine
XS2D_Panel
XS2D_MajorGrid
XS2D_MinorGrid
XS2D_PanelDivider
Geology
GeologyArea
GeologyPoint
GeologyLine
GeologyArea
GeologyLine
GeologyPoint

Introduction

DAVID R. MAIDMENT

Why Arc Hydro for groundwater?

FRESHWATER IS ONE OF THE EARTH'S MOST PRECIOUS RESOURCES, VITAL TO human life and to plant and animal life. The volume of groundwater is much larger than the volume of water in rivers and lakes, but it lies hidden beneath the land surface in aquifers. About 90 percent of the earth's readily available freshwater is stored as groundwater, slowly percolating through the pore spaces and fractures of rocks near the earth's surface. Large regions of the world depend on groundwater for their water supply, particularly in rural areas, because aquifers are spatially extensive.

Groundwater has not been a traditional area of application of geographic information systems (GIS) to water resources, in part because groundwater resources are less visible and readily mapped compared to streams, rivers, lakes, and watersheds. Also, groundwater is inherently a 3D phenomenon because the depth at which water is found in a well is a critical measure of its accessibility. As a well is drilled, its borehole passes through many subsurface strata laid down in layers over geologic time. The spatial extent of these layers is much larger than is their vertical thickness, much like sheets of paper, so 2D GIS mapping of well locations and aquifer boundaries is the normal point of departure for groundwater projects. The core framework of the Arc Hydro data model supports 2D mapping of groundwater resources, and the extensions of this model support 3D representations of boreholes and hydrogeologic strata.

We use a number of terms to define different aspects of Arc Hydro. The name Arc Hydro refers to the overall data model for representing hydrology, including surface water and groundwater. Within this book we refer to this general model as the Arc Hydro data model, or simply as Arc Hydro. The first version of Arc Hydro mainly described surface water systems including drainage, river networks, and time series. This work was published in a book titled *Arc Hydro: GIS for Water Resources* (Maidment 2002). The original Arc Hydro has been redesigned to include a simplified framework for representing the basic surface water and groundwater features. Within this book we refer to the framework data model as the Arc Hydro Framework, or simply as the framework. You can add surface water and groundwater components to the Arc Hydro Framework to describe different aspects of hydrologic systems (the framework data model and the components are described in chapter 2). We refer to the groundwater components of the data model as Arc Hydro Groundwater, or simply as the groundwater data model, which is the focus of *Arc Hydro Groundwater: GIS for Hydrogeology*.

Arc Hydro is not just a set of data models. Rich sets of tools have been developed to help ArcGIS users implement the data model, organize their data, and create GIS products. These tools make use of the standard data structures and the relationships between features within the data model to enhance the capabilities of ArcGIS for the management and analysis of water resources. For surface water analysis, a set of tools named Arc Hydro Tools has been developed by Esri. These tools are available on the Esri Web site (`www.esri.com/archydro`). For groundwater analysis a set of

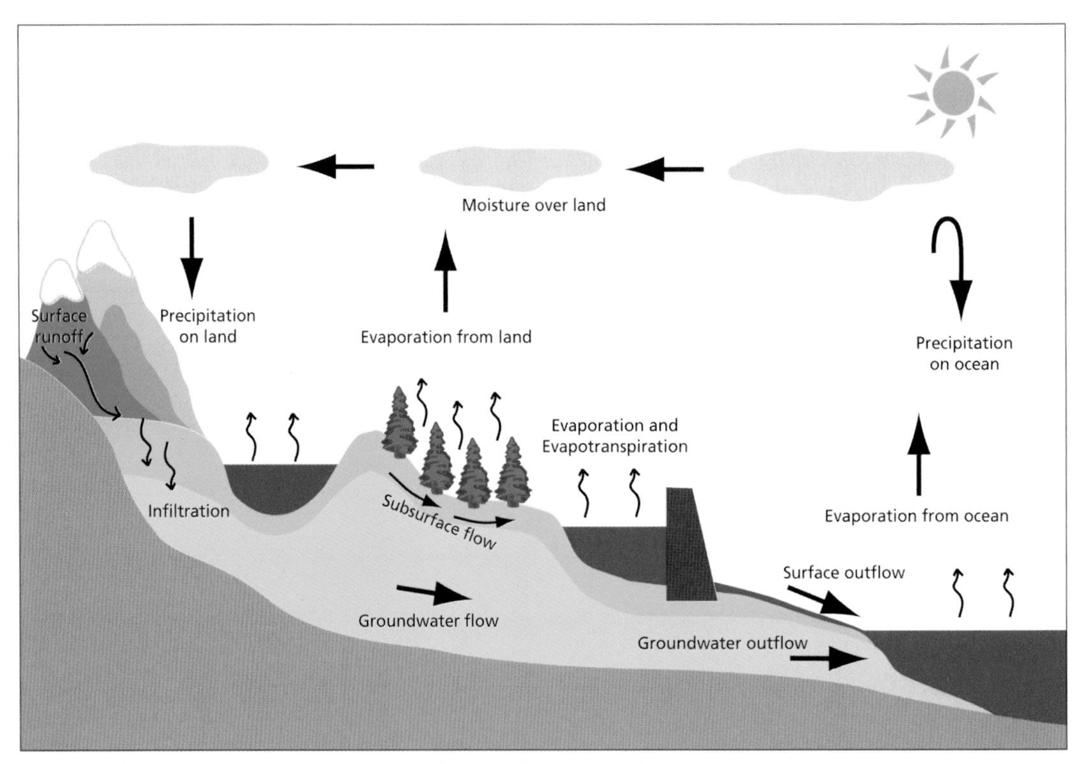

Figure 1.1 The hydrologic cycle depicts the circulation of the waters of the earth.

tools named Arc Hydro Groundwater Tools has been developed by collaboration between Aquaveo, a water resources engineering consulting firm in Provo, Utah, and Esri, and are available on the Aquaveo Web site (`www.aquaveo.com/archydro`).

What makes water different when it comes to GIS?

GIS is applied to many fields of endeavor, indeed to any field in which data can be depicted geospatially. So, what makes water different? Why do we need special geographic data models for water? First of all, water is a subtle substance. It flows from one place to another on the land surface and through the subsurface; water evaporates and travels great distances rapidly in the atmosphere and then returns to the earth again as precipitation. It flows in streams and rivers and accumulates in lakes, bays, and estuaries to form the blue features on topographic maps. Water movement through the hydrologic cycle (figure 1.1) is extremely complex and is still not completely understood.

Using GIS to describe natural water systems requires a means to describe the connectivity of water flow through the landscape. It is not enough to know that there are geographic data layers of water features like streams, aquifers, and wells. It also matters which streams contribute water to a particular aquifer and which wells are drilled into that aquifer to supply water for domestic consumption or irrigation. Only in this way can we understand the inflow and outflow of water, in particular groundwater systems, and thus manage these systems wisely. This book, *Arc Hydro Groundwater: GIS for Hydrogeology*, describes a geographic data model for hydrogeology, using the Edwards Aquifer in Texas as a vehicle for developing and explaining concepts. The surface water components are described in the book *Arc Hydro: GIS for Water Resources* (Maidment 2002), which is planned to be updated to reflect the new Arc Hydro design. These books describe something more than geographic data models for surface and groundwater—they describe a geographic data framework for water, with particular components to depict different aspects of surface and groundwater systems.

Edwards Aquifer

Aquifer systems are vital to the regions that overlie them. The Edwards Aquifer (figure 1.2) underlies the cities of San Antonio, San Marcos, and Austin in south-central Texas. San Antonio is the largest city in the United States whose water supply comes principally from groundwater, distributed throughout the city after being pumped from hundreds of wells drilled into the Edwards Aquifer. Water from these wells supports a population of more than 2 million people.

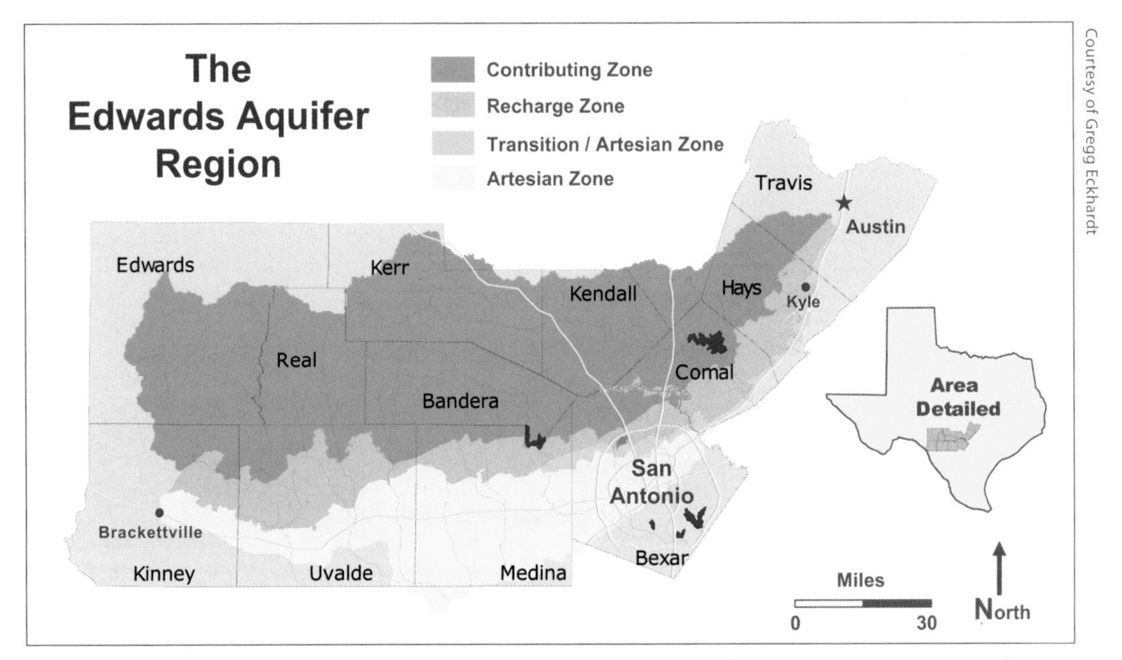

Figure 1.2 Map of the Edwards Aquifer region showing the zones within the aquifer and the recharge zone contributing surface water to the aquifer.

In San Marcos and Austin, beautiful springs discharge water from the Edwards Aquifer. Figure 1.3 shows the pool overlying San Marcos Springs. The deep water is perfectly clear—you can even see sand grains jumping on the bedrock as the groundwater discharges over them. This peaceful haven in San Marcos has sustained people for thousands of years and is considered to be one of the oldest continuously inhabited places in North America.

The Edwards Aquifer is not just an environment of water and rocks. It is home to creatures such as the blind salamander, an almost transparent amphibian perfectly adapted to its life in the perpetual darkness of the aquifer's underground cavities. Indeed, it is the necessity to sustain the spring flow and support the populations of endangered species such as the blind salamander that triggers restrictions on water pumping from the aquifer when water levels are low. The pressure of population growth bears down upon the Edwards Aquifer. New housing developments, shops, offices, and roads are spreading where once there was a quiet forest. Bacteria and chemicals washed down with runoff waters are invading the pristine aquifer environment. Careful management of both water quantity and quality in the aquifer are needed to preserve this precious asset to sustain future generations as it has sustained those of the past.

Courtesy of Gregg Eckhardt

Figure 1.3 Tourists get a picturesque view of the abundant plant and fish life in the San Marcos Springs through a glass-bottom boat.

Surface water and groundwater

Groundwater is strongly related with surface water in the Edwards Aquifer. Water seeping through the limestone rock of the aquifer has created a karst landscape with many caves, fractures, and sinkholes that allow fast movement of water from the surface into the aquifer formation and within the subsurface (figure 1.4).

The groundwater discharge from the San Marcos Springs forms the flow of the San Marcos River. As this river meanders downstream, it is joined by the Blanco River, whose flow is derived mostly from surface runoff, and these in turn are tributaries of the Guadalupe River, which carries their waters to the Gulf of Mexico. Normally a quiet, slow-moving river, the Guadalupe can turn into a raging torrent during severe storms and devastate the surrounding countryside.

To accurately describe the water resources of the region, we must map both groundwater and surface water features together (figure 1.5) and define the relationships between them. This is part of the new design of Arc Hydro, which includes a framework data model with basic surface water and groundwater features and additional components describing unique aspects of surface water and groundwater.

Courtesy of the Edwards Aquifer Authority

Figure 1.4 On top is a limestone bedrock with dissolution features. Over time, water slowly dissolves the rock and creates fractures and recharge features. The bottom figure shows a 70-foot-deep vertical shaft recharging water into the aquifer.

Groundwater datasets

In the United States, data depicted on blue lines, the color of surface water features on topographic maps, have been compiled by the Unites States Geological Survey (USGS) into seamless national hydrography datasets at scales of 1:100,000 and 1:24,000, but no equivalent national hydrogeology dataset exists. Subsurface hydrogeologic data are measured and archived by many federal, state, and local groundwater agencies in a fragmented way without a common means of data access and synthesis. The lack of a systematic organization of hydrogeologic and groundwater data means that their formats vary from state to state, from location to location, and from project to project. A new groundwater investigation can be like an Easter egg hunt, where you search around for the basic data needed to support the investigation, coping with many disparate data types and formats from different data sources.

We envisage that the adoption of the Arc Hydro Groundwater data model will lead to better organization of groundwater data so that standardized groundwater and hydrogeologic datasets can be systematically compiled and maintained. Since hydrogeology is a subfield of geology, it is appropriate that Arc Hydro Groundwater has a geology component to allow for incorporating existing geologic maps and data into Arc Hydro Groundwater datasets.

The ArcGIS Desktop software system is the key means by which groundwater information is compiled and synthesized in Arc Hydro Groundwater. However, this is now being supplemented by ArcGIS Server and ArcGIS Online, by which compiled Arc Hydro datasets can be published on the Internet as maps and data services. This emerging "services-oriented architecture" for geospatial information is an important way for separate organizations to publish geographically distributed groundwater datasets in a manner that facilitates their assembly and integration over a region.

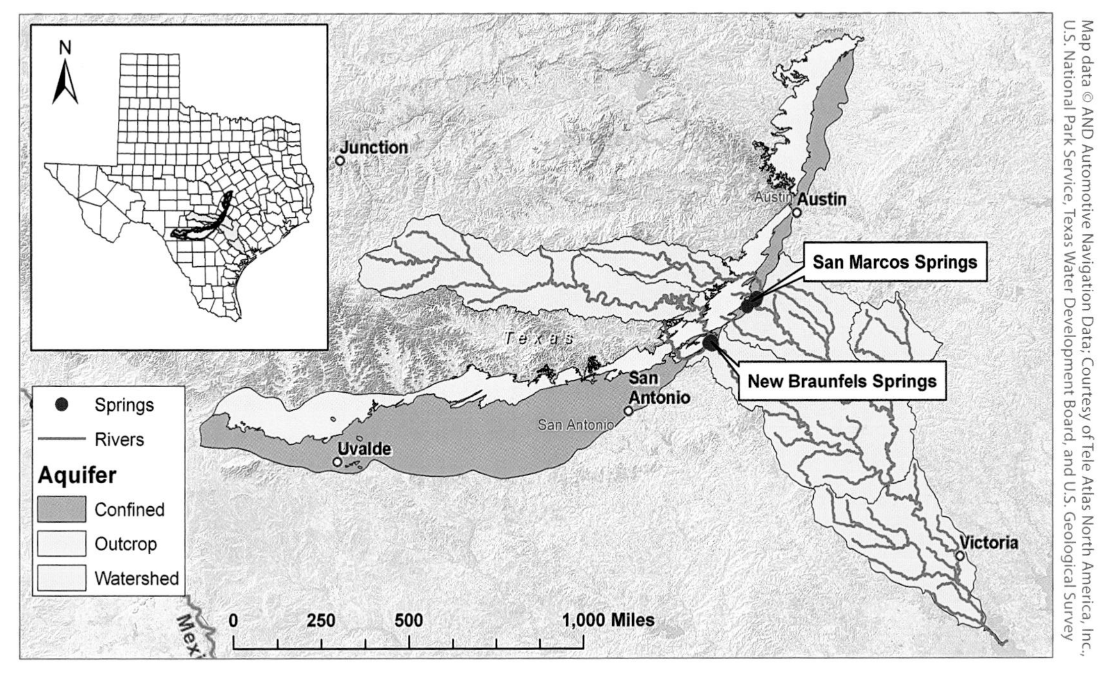

Figure 1.5 The Edwards Aquifer (in brown) and the Guadalupe River Basin (with the watershed in green and rivers in blue) are located in Central Texas. A strong connectivity exists between surface water recharging the aquifer through streams flowing over the outcrop of the aquifer and groundwater discharging through springs.

Groundwater wells

Wells provide the most abundant source of groundwater information. Many large tabular databases describing wells exist, both as "well logs" that describe the details of well construction and the subsurface materials encountered while drilling, and as "observations" that describe the water level and sometimes the chemical properties of the well water indexed by time. This information is sparse both in space and in time: sparse in space because each well is an individual entity, and they are drilled relatively far apart from one another; sparse in time because groundwater conditions change slowly, and most wells are manually sampled infrequently over periods of months or years. Figure 1.6 shows an artesian groundwater well, with water flowing freely from the subsurface. This is common to the artesian zones of the Edwards Aquifer.

This sparse information must be interpreted and interpolated to generate spatial maps of groundwater properties, in particular to produce piezometric head maps that describe the spatial patterns of groundwater levels (or pressures) in aquifers. If piezometric head maps are constructed for successive periods of time and subtracted one from the next, a picture of the rise and fall of groundwater conditions through time can be obtained. The Arc Hydro Groundwater tools include geoprocessing tools and models for accomplishing this task (see chapter 7).

Courtesy of the Edwards Aquifer Authority

Figure 1.6 Groundwater under pressure flows freely from a confined aquifer to the surface of this flowing artesian well.

Groundwater strata

Understanding groundwater systems requires being able to depict their horizontal extent and also their vertical structure, usually visualized as cross-sections cut through the various hydrogeologic strata (figure 1.7).

In Arc Hydro Groundwater, tabular information from well logs about the "contact points" at which the well driller encountered each new subsurface layer is translated within ArcGIS into a 3D vertical representation of the borehole. Cross-sections can be drawn by connecting a sequence of boreholes or by cutting directly through a "geovolume" representation of the 3D structure of these hydrogeologic units. A special extension of the Arc Hydro Groundwater tools called Subsurface Analyst can be used to perform these functions.

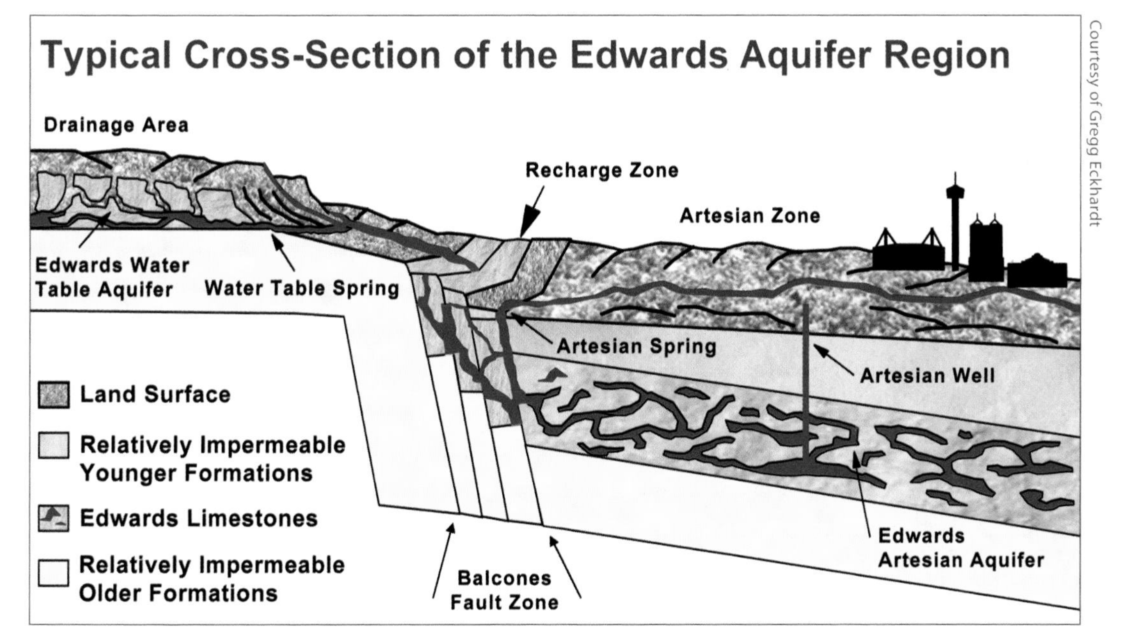

Figure 1.7 A typical cross section of the Edwards Aquifer describes strata, hydrogeologic units, and features such as faults and wells within the subsurface.

Groundwater modeling

Because groundwater systems are largely hidden from sight, groundwater simulation models are used to infer the properties of groundwater systems and to simulate the flow and storage of water within the subsurface under variable conditions, including natural conditions and human-made influences such as pumping. The most widely used groundwater simulation model is MODFLOW, produced by the USGS, which is as much of a standard in the groundwater field as ArcGIS is a standard for geographic information systems.

The various data arrays in a MODFLOW model are like the pages in a groundwater system book. As you view them one by one, you see the spatial extent and physical properties of the aquifer, such as its thickness, porosity, and hydraulic conductivity (figure 1.8). You see the pumping stresses, flow patterns, and recharge and discharge volumes, and you see the piezometric head map that depicts the levels of water within the different layers of the groundwater system.

The Arc Hydro Groundwater tools include a special extension called MODFLOW Analyst that enables MODFLOW descriptions of groundwater

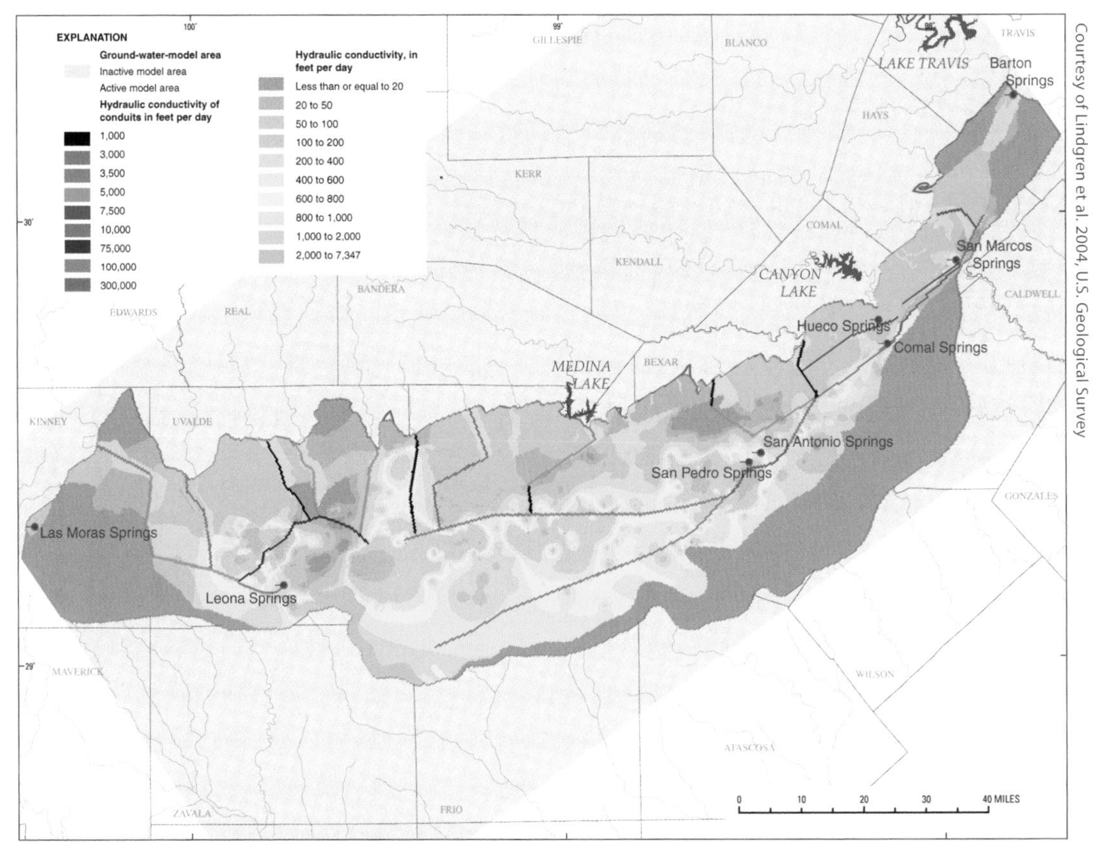

Courtesy of Lindgren et al. 2004, U.S. Geological Survey

Figure 1.8 Simulated distribution of horizontal hydraulic conductivity for a calibrated MODFLOW model of the southern part of the Edwards Aquifer. Warmer colors represent higher conductivity and faster movement of water within the aquifer. The linear features within the model represent conduits with very high conductivity.

systems to be viewed and mapped in ArcGIS, thus enabling a much larger audience to understand the behavior of groundwater systems simulated with MODFLOW (see chapter 8).

Scope of the book

The book is divided into nine chapters. The first three chapters introduce the subject, describe the core framework data model, and present an exposition of how to represent 3D objects in ArcGIS. The next three chapters present components of the groundwater data model for geological mapping, wells and aquifers, and 3D hydrostratigraphy. The final three chapters describe time series, groundwater modeling, and implementation of the groundwater data model in ArcGIS.

It is important to understand some of the limitations of this book—we are describing an information model for groundwater data. This is not a book about groundwater simulation modeling. Arc Hydro Groundwater can be used to help create groundwater simulation models and to store the results of groundwater simulations, but Arc Hydro Groundwater is not itself a groundwater simulation model. Likewise, although we devote considerable attention to hydrogeology, Arc Hydro Groundwater is not intended to capture all the detailed cartographic aspects of geologic and hydrogeologic mapping. Our focus is on the groundwater environment and on the properties of the water flowing through that environment.

References

Lindgren, R. J., A. R. Dutton, S. D. Hovorka, S. R. H. Worthington, and P. Scott. 2004. Conceptualization and simulation of the Edwards aquifer, San Antonio region, Texas: U.S. Geological Survey Scientific Investigations Report 2004–5277, 143 p.

Maidment, D. R., ed. 2002. *Arc Hydro: GIS for Water Resources*. Redlands, California: Esri Press.

2

chapter two

Arc Hydro framework

DAVID R. MAIDMENT AND GIL STRASSBERG

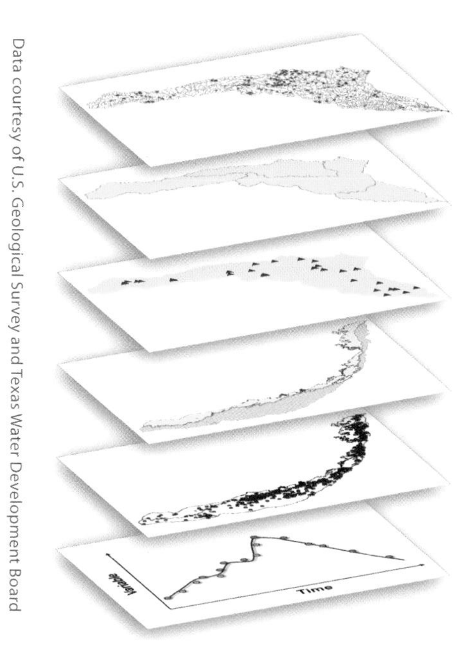

Data courtesy of U.S. Geological Survey and Texas Water Development Board

Thematic layers of the Arc Hydro framework

Hydrography (WaterLine, Waterbody, and WaterPoint)
The "blue lines" on maps representing rivers and streams, water bodies, and points of interest such as springs and diversions points.
Representation: Points, Lines, and Polygon features

Watershed
Drainage areas defining the extent of land from which water drains into a body of water.
Representation: Polygon features

MonitoringPoint
Locations where water is measured, such as stream and precipitation gages, and water quality monitoring stations.
Representation: Point features

Aquifer
Geologic formations that yield significant quantities of water to springs and wells.
Representation: Polygon features

Well
Man-made structure to withdraw or sample groundwater.
Representation: Point features

Time Series
Temporal information describing water quantity and properties.
Representation: Tabular data related to spatial features

WHEN A STUDY IS UNDERTAKEN USING ARC Hydro Groundwater, the first step is to compile readily available data into a geodatabase so that they form a foundation upon which more detailed datasets and interpretations can later be built. The Arc Hydro Framework is formed of feature classes representing common surface and groundwater features, such as hydrography, watersheds, monitoring points, aquifers, wells, and time series of water observations data. The same Arc Hydro Framework datasets are used for surface water studies implementing Arc Hydro as well, so that we can think of water as a single resource. In a particular study, some feature classes or tables of the Arc Hydro Framework are not needed; those can be retained in the framework data structure but not filled with data.

Once the Arc Hydro Framework is created, other data model components can be added, as shown in figure 2.1. In the case of groundwater, these include components for geological maps, 3D description of boreholes and hydrostratigraphy, and links to groundwater simulation models. In the case of surface water, the additional components include a GIS network representation of the stream system, catchments, and watersheds that convey drainage to the stream network, description of hydrography, including rivers and water bodies (e.g., lakes, estuaries, marshes and bays), and 3D representation of river channels. Both Arc Hydro surface water and groundwater share a common description of temporal data such as time series of water levels and water quality measured in wells.

An important design consideration for the Arc Hydro data model is to have a parsimonious data model—namely to limit the number of required feature classes, tables, and attributes to the minimum needed to achieve a description of the basics of a groundwater system. Often, more information is required than this basic description provides. Indeed, the Arc Hydro Groundwater tools themselves generate additional tables and attributes shown in this book. The data model presented here is intended to be a basis for further elaboration in your context rather than an exhaustive description of every possible variant for describing groundwater systems.

We created the groundwater data model to focus on classes of information for which data are readily available in most applications and to

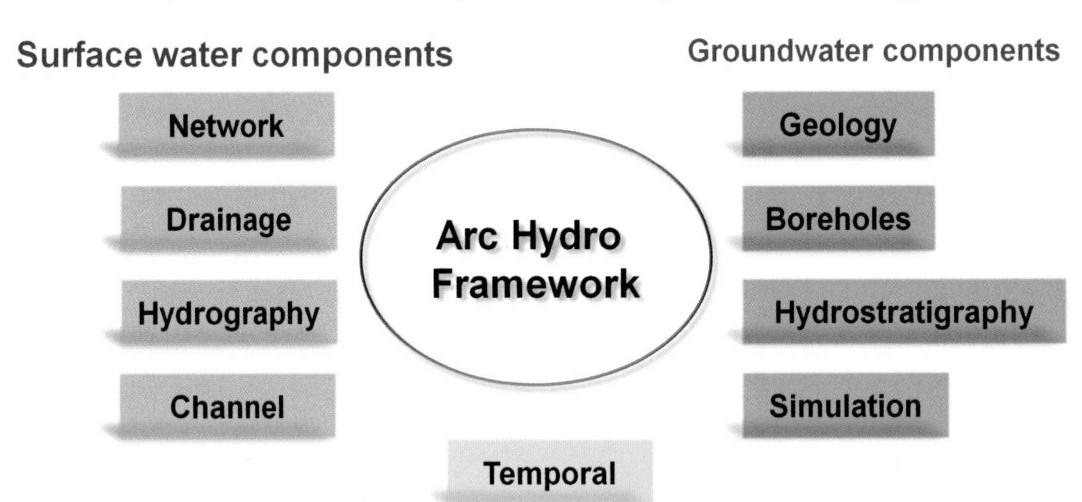

Figure 2.1 Components of the Arc Hydro data model (including surface water and groundwater).

support implementation workflows that create useful information products. As such, each chapter that outlines a component of the groundwater data model concludes by describing the process of implementing that component in a useful way.

The Arc Hydro Groundwater data model components are described further in chapters 4 through 8 of this book:

- Geology—Point, line, and area features from geologic maps;
- Aquifer, well, and borehole data—2D representation of aquifers and wells, and the borehole component that represents 3D data along boreholes;
- Hydrostratigraphy—Representation of hydrogeologic units with 2D and 3D features (cross-sections and geovolumes) and surfaces;
- Simulation—Groundwater simulation models represented as 2D and 3D features;
- Temporal—Representation of spatial-temporal datasets.

As noted in chapter 1, an update to *Arc Hydro: GIS for Water Resources* (Maidment 2002) is envisioned to better describe the components of a data model for surface water.

Desktop, server, and online GIS

When *Arc Hydro: GIS for Water Resources* was published in 2002, it was conceived as a data model to be used within ArcGIS Desktop to ingest and process information from various local sources. More than a hundred Arc Hydro tools have been built on this foundation in subsequent years, and the Arc Hydro toolset is now the most widely used means of performing tasks such as watershed and stream network delineation from digital elevation models (DEMs). Very often the input data for such a process are downloaded from the Internet, such as getting a DEM dataset for a study area from the USGS Seamless Server.

In subsequent years, other patterns for using geospatial information have emerged, including ArcGIS Server and ArcGIS Online. ArcGIS Server is a mechanism for publishing information as geographic data services and tiled image maps that can be rapidly ingested through the Web into ArcGIS Desktop or other applications such as mobile GIS, or Web interfaces for geospatial data query and analysis.

ArcGIS Online is a central repository of information in "the cloud" that is directly accessible through Web connections to ArcGIS Desktop, and it supplies global coverage of high-quality basemaps for topography, roads, hydrography, geology, orthophotography, and other themes. These 2D basemaps are tiled for rapid display at any spatial scale, and they provide important context for the detailed feature representations created in Arc Hydro Groundwater. New information, such as a completed Arc Hydro Groundwater geodatabase, can be symbolized in an ArcMap display, assembled into a "Layer Package," and published in ArcGIS Online, either publicly or shared just within a group of other ArcGIS Online users. In this sense, ArcGIS Online is a Web-based social network for map and data sharing.

The emergence of ArcGIS Server and ArcGIS Online as automated mechanisms for map and data sharing means that creating Arc Hydro geodatabases is a more valuable activity than was formerly the case, because the constructed result can more readily be accessed and used by others.

The importance of these developments in Web access to geospatial information should not be underestimated. The United States has a National Hydrography Dataset for surface water features but lacks a National Hydrogeologic Dataset for groundwater features. Thus, Web-based technology now exists to develop and publish regionally consistent hydrogeologic datasets even when no national system is available.

Time series of water observations data such as streamflow and groundwater levels are now being published by the U.S. Geological Survey and other water agencies as Web services in the WaterML language. WaterML is an eXtended Markup Language (XML) specification that describes the site at which measurements are made, such as a well; the variables measured there; the time period of record and number of values of each variable available; and the time series of the measured values themselves. WaterML Web services can be used to directly ingest water observations data into the Arc Hydro temporal component, where Arc Hydro Groundwater tools can be used to create time varying maps of groundwater conditions such as piezometric head levels or concentrations of a chemical constituent.

The simultaneous advent of Web services for water observations data such as groundwater levels and water quality means that the geographic data services just described can be augmented by water observations data services supplying current information about water conditions in aquifer systems. These GIS and water observation data services taken together constitute the basis for implementing a services-oriented architecture for groundwater information, a concept from computer science that promises to greatly facilitate the synthesis of otherwise fragmented groundwater information housed at water agencies in different geographic locations. The process for publishing and ingesting Arc Hydro Groundwater data services is explained in chapter 9.

Building a geographic data model

Building a geographic data model requires some general principles. In part, these are derived from the nature of ArcGIS itself, but in the larger sense, geographic data modeling is part of a wider discipline of representing the world by abstract models. Booch et al. (1999) define a model as a "simplification of reality" aimed to better understand the system we are studying. Data models help us describe complex systems using a structured set of data objects. A geographic data model is a representation of the real world expressed with spatial datasets within a GIS.

Geographic data models are useful for two major reasons:

1. They provide a template for modeling types of systems by defining a set of common spatial objects and the relationships between them.

2. They define a common vocabulary to assist sharing of data, tools, and analyses within a discipline.

Designing a geographic data model includes three phases: conceptual design, logical design, and physical design (Arctur and Zeiler 2004). The conceptual design includes the identification of information products created with GIS and defines the key thematic layers and how these can be grouped into datasets. The logical design defines the structure and behavior of the datasets and results in a geodatabase prototype. The physical design includes the implementation process where the prototype is tested through case studies and reviewed by users. This is an iterative process of refining the data model based on reviews and results from case studies. During the logical design we define a set of thematic layers that describe objects in the modeled system. Thematic layers are defined by classifying and grouping objects with similar geometries, properties, and behavior. For example, in the groundwater data model, all wells (irrigation, domestic, industrial, etc.) are grouped into one thematic layer and are described as point features with similar attributes.

The behavior of the system, or "how the system works," is described by relationships between objects of the data model. Feature-to-feature relationships enable us to track the movement of water through the hydrologic system. Water can be traced from where it drops on the land surface to the outflow point of the watershed in which it fell and through the downstream water bodies and monitoring points on its way to the ocean (figure 2.2). In some cases water can be traced through the stream network and over an aquifer outcrop where water can enter the underling aquifer, ultimately flowing to a well within the aquifer or discharging at a spring. The ability to create relationships between spatial datasets is an important capability provided as part of the geodatabase environment.

Geodatabases

A geodatabase is a repository of geographic information organized into geographic datasets. It provides a common data storage and management framework for storing all the types of datasets supported in ArcGIS. There are three types of geodatabases: personal geodatabases, file geodatabases, and ArcSDE geodatabases. The personal and ArcSDE geodatabases are built on top of relational database management systems (RDBMS) such as Microsoft Access, Oracle, or Microsoft SQL server, which are customized for storing spatial data structures. The file geodatabase manages data in a

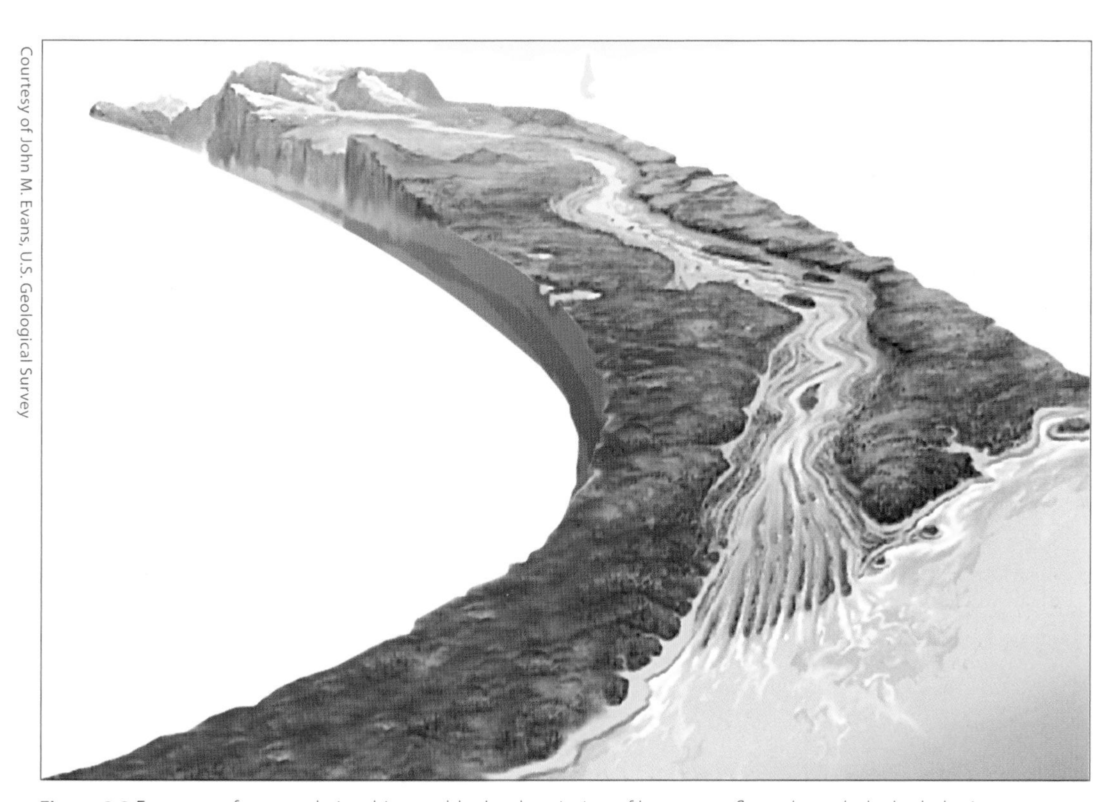

Courtesy of John M. Evans, U.S. Geological Survey

Figure 2.2 Feature-to-feature relationships enable the description of how water flows through the hydrologic system.

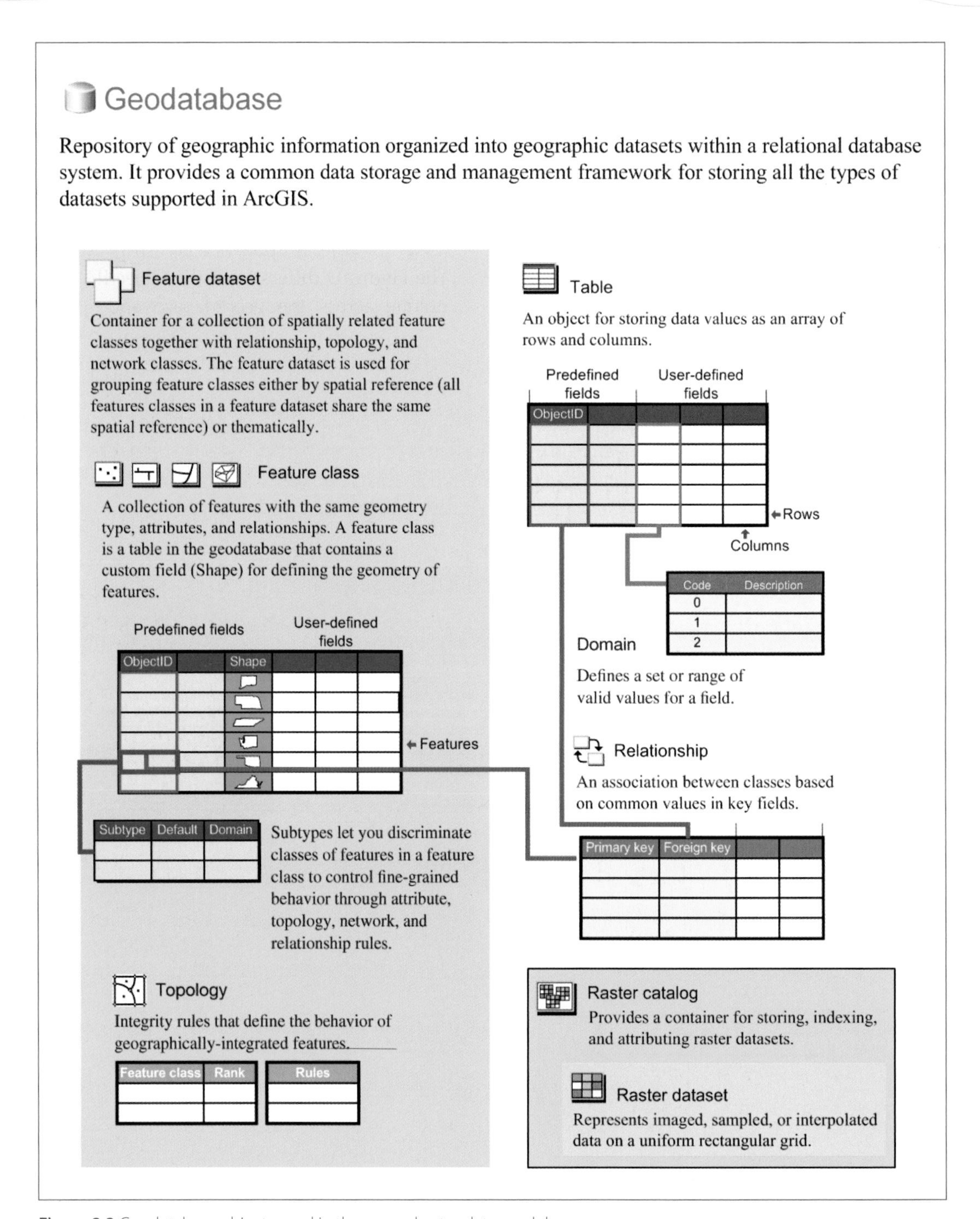

Figure 2.3 Geodatabase objects used in the groundwater data model.

file system folder. For a more detailed description of the types of geodatabase and the differences between them see the ArcGIS desktop help system (`http://webhelp.esri.com`). Storing data in a geodatabase has a number of technical advantages. A geodatabase can store a variety of spatial and tabular datasets in one centralized location, support versioning and multiuser access to the data, and allow easy scaling of storage solutions.

Data objects in the Arc Hydro Groundwater geodatabase include feature classes, feature datasets, raster datasets, raster catalogs, tables, and relationships (figure 2.3). A more detailed description of geodatabase objects is provided by Zeiler (1999).

Hydro features

As part of the Arc Hydro design, generic feature classes are customized by adding HydroID and HydroCode attributes to them. We call these modified features hydro features. Generic geodatabase objects contain an ObjectID field, which is maintained by ArcGIS and guarantees a unique ID for each row in a table (also for features in a feature class and rasters in a raster catalog). The two custom attributes, HydroID and HydroCode, are added to spatial features in the Arc Hydro geodatabase to provide a mechanism for uniquely identifying features across the geodatabase, establishing relationships between classes, and tracing data back to its source in external information systems (figure 2.4).

- HydroID is an integer attribute that uniquely identifies objects in an Arc Hydro geodatabase. The HydroID differs from the ObjectID, as it is unique across the geodatabase and not only within a certain class (table, feature class, raster catalog).

- HydroCode is a text attribute that is a permanent public identifier of a feature. The HydroCode provides a linkage with external information systems.

For example, in chapter 7 we describe how wells (described as point features) are related with water-level measurements stored on the USGS National Water Information System (NWIS). The HydroID of the well feature is unique within the geodatabase, but the USGS system also has its own unique identifier. In this case the well site number of 295443097554201 uniquely identifies this well from all other wells in the USGS system.

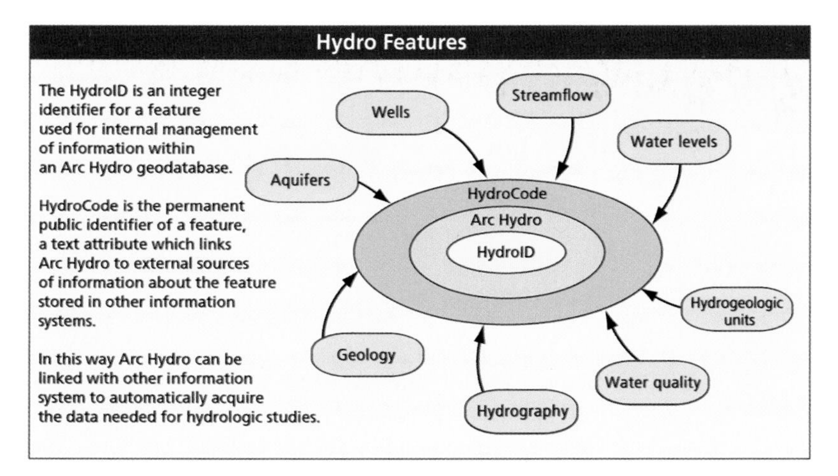

Figure 2.4 Hydro features are created by adding HydroID and HydroCode attributes to spatial features in an Arc Hydro geodatabase.

The well site number is stored in the HydroCode attribute of the well feature, which enables querying data (e.g., water levels, water quality) from the USGS external database. Tools developed on top of the Arc Hydro data model enable the retrieval of data from local tables or over the internet based on the HydroCode values. In this way Arc Hydro can be linked with external information systems to acquire data necessary for hydrologic analysis.

HydroID is managed carefully because it is such an important attribute within Arc Hydro. HydroIDs are managed using a pair of tables, named APUNIQUEID and LAYERKEYTABLE, which are created automatically when you use the Arc Hydro tools (figure 2.5). HydroID values are tracked in the APUNIQUEID table. Each time a new ID is assigned to a feature in the geodatabase, a counter is updated so that the same HydroID is never assigned again within the given geodatabase. The APUNIQUEID table includes two fields: IDNAME is the name of the key field (e.g., HydroID), and LASTID stores the last ID used for that key field. The LASTID field is updated accordingly each time new IDs are assigned. An additional table, LAYERKEYTABLE, can be used to manage multiple identifiers, so that data from different sources or for different projects or separate areas can be attributed with separate sets of unique identifiers.

Hydro features in the geodatabase can be associated with other hydro features by storing their HydroID values as an attribute. By this process, river reaches can be associated with aquifers, which can be associated with wells, thus defining the movement of water between the stream and wells. Similarly, time-series data can be associated with hydro features simply by storing the HydroID of a feature as an attribute of the time-series record. Uniquely labeling all hydro features in a geodatabase is a powerful concept for supporting behavioral modeling, because the geodatabase can be considered as an integrated whole rather than a set of separate data layers.

In Arc Hydro, all fields used to support HydroID-based relationships are of type integer. Integers are easier to manage than text strings, and database queries operate more efficiently when integers index data values. All Arc Hydro attributes ending in ID (e.g., HydroID, AquiferID, WellID, and FeatureID) indicate an integer identifier, and attributes ending in Code (e.g., HydroCode) indicate a text identifier.

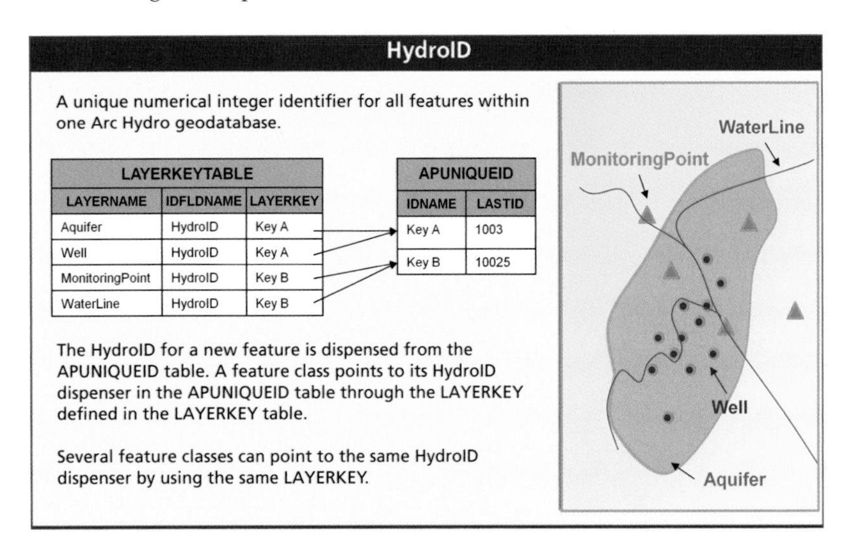

LAYERKEYTABLE		
LAYERNAME	IDFLDNAME	LAYERKEY
Aquifer	HydroID	Key A
Well	HydroID	Key A
MonitoringPoint	HydroID	Key B
WaterLine	HydroID	Key B

APUNIQUEID	
IDNAME	LASTID
Key A	1003
Key B	10025

Figure 2.5 HydroIDs within an Arc Hydro geodatabase are managed using the APUNIQUEID and LAYERKEY tables.

Arc Hydro framework

The Arc Hydro framework (figure 2.6) provides a simple data structure for storing the most basic geospatial datasets for describing hydrologic systems. This framework supports basic water resources analysis such as tracing water as it flows over the terrain in watersheds, streams, and water bodies; creating groundwater level and groundwater quality maps; and viewing time-series data related with monitoring stations and wells. The framework serves as a simple point of departure to which more detailed components can be added, as discussed in later chapters.

The creation of an integrated geodatabase instead of a collection of data layers is a key accomplishment of the Arc Hydro design, providing a starting point for building stronger water resources applications in GIS. The green lines in figure 2.6 show the relationships between the features in the framework. Each relationship has a cardinality (represented by the numbers in the analysis diagram) that defines the number of features in each class that can be related. In the framework data model all relationships are one-to-many (1:M), meaning that a single feature in one class can be related to one or more features in the related class. For example one Aquifer feature can be associated with one or more Well features. The establishment of relationships allows the association between features and objects in the data model so that Well features can be linked with Aquifer features. Although no explicit relationship is built, WaterLine, Waterbody, and WaterPoint features

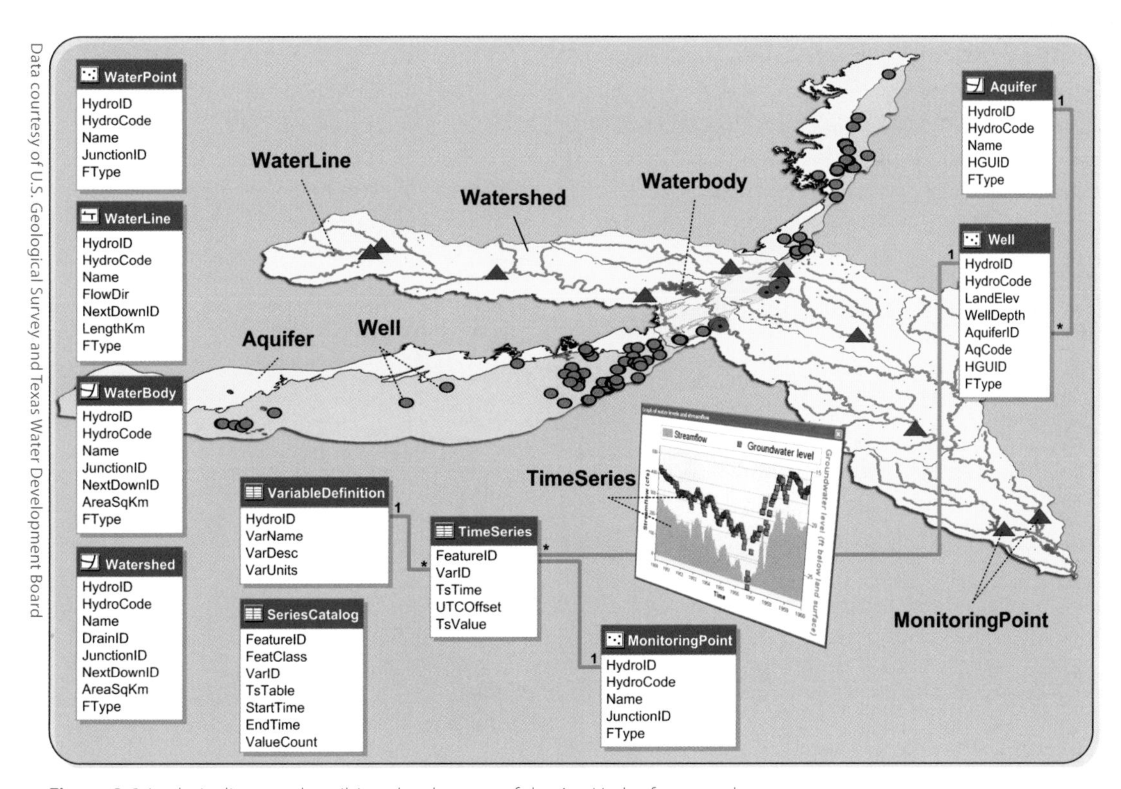

Figure 2.6 Analysis diagram describing the datasets of the Arc Hydro framework.

can also be associated with Aquifer features. Since Aquifer features are related to Well features, one can trace the movement of water from where it falls on the land surface through the drainage and river network and into the subsurface. In addition, MonitoringPoints and Wells are related with time series data to enable storage and visualization of measurements recorded at monitoring stations and wells.

Hydrography (the blue lines on a map) and drainage areas which contribute flow to the water system are represented by a set of feature classes describing river networks, water bodies, watersheds, and special points of interest in the hydrologic system. Feature classes describing hydrography include WaterLine, WaterBody and WaterPoint (figure 2.7). WaterLine is a line feature class for representing hydrographic "blue lines," which represent mapped streams and water body center lines. The features are usually directional (digitized with the flow direction) and connected so that they can form a network. WaterBody is a polygon feature class for representing areal water features in the water system, such as lakes, ponds, swamps, and estuaries. WaterPoint is a point feature class for representing hydrographic features such as springs, water withdrawal/discharge locations, and structures.

WaterLine

Line feature class for storing hydrographic "blue lines" representing mapped streams and water body center lines

Field name	Description
HydroID	Unique feature identifier in the geodatabase used for creating relationships between classes of the data model.
HydroCode	Permanent public identifier of the feature used for relating features with external information systems.
Name	Text attribute representing the name of the river or stream.
FlowDir (flow direction)	Direction of flow for the feature (e.g., with digitized, against digitized).
NextDownID	Relates a WaterLine feature to its downstream feature, thus creating river network connectivity. NextDownID is equal to the HydroID of the next downstream feature.
LengthKm (length in km)	Length of the WaterLine feature in km units. This attribute is commonly used for modeling purposes and analytic calculations.
FType (feature type)	Distinguishes between types of WaterLine features.

Waterbody

Polygon feature class for representing areal water features in the water system

Field name	Description
HydroID	Unique feature identifier in the geodatabase used for creating relationships between classes of the data model.
HydroCode	Permanent public identifier of the feature used for relating features with external information systems.
Name	Text attribute representing the name of the water body.
JunctionID	Relates a Waterbody feature with a river network by associating a water body with a junction on the network. The JunctionID of a Waterbody feature is equal to the HydroID of a related HydroJunction feature.
NextDownID	Relates a WaterBody feature to its downstream feature, thus creating feature to feature connectivity. NextDownID of a WaterBody feature is equal to the HydroID of the next downstream feature.
AreaSqKm (area in square km)	Area of the water body in square km units. This attribute is commonly used for modeling purposes and analytic calculations.
FType (feature type)	Distinguish between types of water bodies (e.g., lake, pond, and estuary).

WaterPoint

Point feature class for representing hydrographic features

Field name	Description
HydroID	Unique feature identifier in the geodatabase used for creating relationships between classes of the data model.
HydroCode	Permanent public identifier of the feature used for relating features with external information systems.
Name	Text attribute representing the name of the water point.
JunctionID	Relates a HydroPoint feature with a river network by associating the point feature with a junction on the network. The JunctionID of a HydroPoint feature is equal to the HydroID of a related HydroJunction feature.
FType (feature type)	Distinguishes between types of HydroPoint features (e.g., spring, diversion point, structure).

Figure 2.7 WaterLine, Waterbody, and WaterPoint feature classes represent hydrography—the blue lines on a map.

Watershed is a polygon feature class for representing drainage areas contributing flow from the land surface to the water system (figure 2.8).

Time series are measured at different locations and represent different types of water properties and quantities. There are many types of monitoring points, such as stream gages where flow is measured, precipitation gages, and sampling locations where water properties (e.g., temperature, pH, and water quality) are measured. MonitoringPoint is a point feature class for representing locations where water is measured (figure 2.9).

A number of the framework feature classes include a JunctionID attribute that relates these features to a river network. The river network itself is included as a separate component, "Network," which is based on a geometric network dataset.

The most basic features commonly used to represent groundwater systems are aquifers and wells. Aquifers are commonly delineated in aquifer maps that describe the boundaries of aquifers and zones within the aquifers such as unconfined and confined sections. Aquifer is a polygon feature class, where each feature represents an aquifer or a part of an aquifer (figure 2.10).

Watershed

Polygon feature class for representing drainage areas contributing flow from the land surface to the water system

Field name	Description
HydroID	Unique feature identifier in the geodatabase used for creating relationships between classes of the data model.
HydroCode	Permanent public identifier of the feature used for relating features with external information systems.
Name	Text attribute representing the name of the watershed.
DrainID	Index associating Watershed features with a specific drainage area. The DrainID of a Watershed feature is equal to the HydroID of the reference drainage area feature.
JunctionID	Relates a Watershed feature with a river network by associating a watershed with a junction on the network. The JunctionID of a Watershed feature is equal to the HydroID of a related HydroJunction feature.
NextDownID	Relates a Watershed feature to its downstream feature, thus creating feature to feature connectivity. NextDownID is equal to the HydroID of the next downstream feature.
AreaSqKm (area in square km)	Area of the watershed in square km units. This attribute is commonly used for modeling purposes and analytic calculations.
FType (feature type)	Distinguishes between types of Watershed features.

Figure 2.8 Watershed feature class for representing drainage areas.

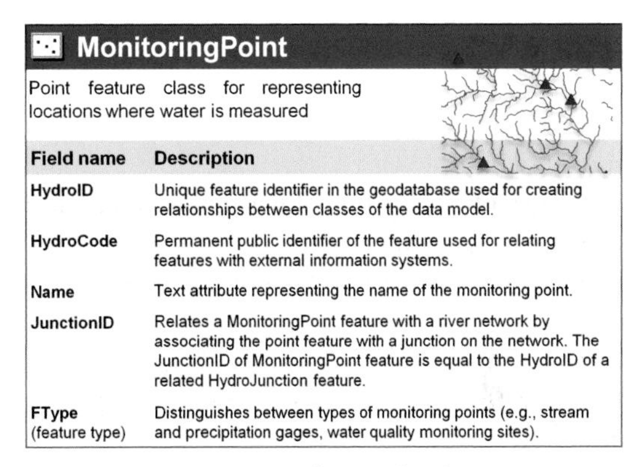

MonitoringPoint

Point feature class for representing locations where water is measured

Field name	Description
HydroID	Unique feature identifier in the geodatabase used for creating relationships between classes of the data model.
HydroCode	Permanent public identifier of the feature used for relating features with external information systems.
Name	Text attribute representing the name of the monitoring point.
JunctionID	Relates a MonitoringPoint feature with a river network by associating the point feature with a junction on the network. The JunctionID of MonitoringPoint feature is equal to the HydroID of a related HydroJunction feature.
FType (feature type)	Distinguishes between types of monitoring points (e.g., stream and precipitation gages, water quality monitoring sites).

Figure 2.9 MonitoringPoint feature class for representing locations where water is measured.

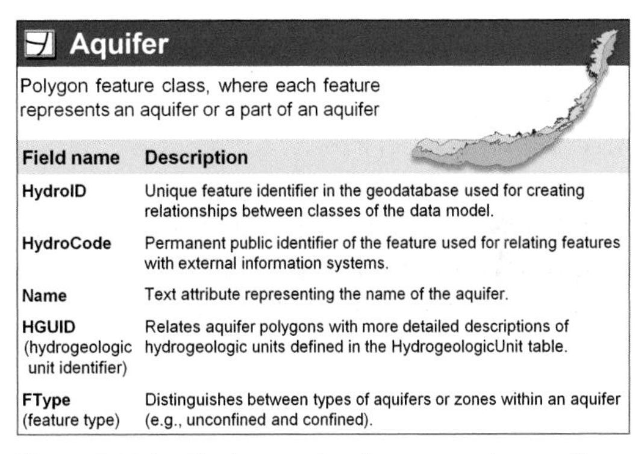

Aquifer

Polygon feature class, where each feature represents an aquifer or a part of an aquifer

Field name	Description
HydroID	Unique feature identifier in the geodatabase used for creating relationships between classes of the data model.
HydroCode	Permanent public identifier of the feature used for relating features with external information systems.
Name	Text attribute representing the name of the aquifer.
HGUID (hydrogeologic unit identifier)	Relates aquifer polygons with more detailed descriptions of hydrogeologic units defined in the HydrogeologicUnit table.
FType (feature type)	Distinguishes between types of aquifers or zones within an aquifer (e.g., unconfined and confined).

Figure 2.10 Aquifer feature class for representing aquifer boundaries.

Wells are commonly described in groundwater databases, and most of the transient data collected for describing the variability in groundwater quantity and quality are collected at wells. Well is a point feature class for representing well locations and their attributes (figure 2.11).

Transient data describing the variation in flow, water levels, precipitation, and water quality are commonly collected at monitoring points and at wells. The Arc Hydro framework includes a simple tabular structure to store such transient measurements, to index different types of time series, and to relate the time series with spatial features. The Arc Hydro framework uses relationships to associate time series with monitoring points and wells. These relationships are a one-to-many type, thus a monitoring point or well feature can have many time series records (of different types).

The temporal component of the framework includes three tables: TimeSeries, VariableDefinition, and SeriesCatalog (figure 2.12). TimeSeries is a tabular dataset for storing time series values indexed by location, time, and the type of the

Well

Point feature class for representing well locations and their attributes.

Field name	Description
HydroID	Unique feature identifier in the geodatabase used for creating relationships between classes of the data model.
HydroCode	Permanent public identifier of the feature used for relating features with external information systems.
LandElev (land elevation)	The elevation of the land surface at the well location. Is commonly used to reference vertical information (measured as depth along the well).
WellDepth (well depth)	The depth of the well. Together with the LandElev provides a description of the well's 3D geometry.
AquiferID (aquifer identifier)	Relates a Well feature with an Aquifer feature. The AquiferID of a Well feature is equal to the HydroID of an Aquifer feature.
AqCode (aquifer Code)	Text describing the aquifer. Is used to symbolize wells based on the related aquifer.
HGUID (hydrogeologic unit identifier)	Relates the well to a hydrogeologic unit.
FType (feature type)	Distinguishes between types of wells (e.g., domestic, water supply, industrial).

Figure 2.11 Well feature class for representing the location of wells and their basic attributes.

time-series data, VariableDefinition is a table with attributes to describe the nature of a variable represented in the TimeSeries table, and SeriesCatalog is a table for indexing and summarizing time series stored in the TimeSeries table. A more

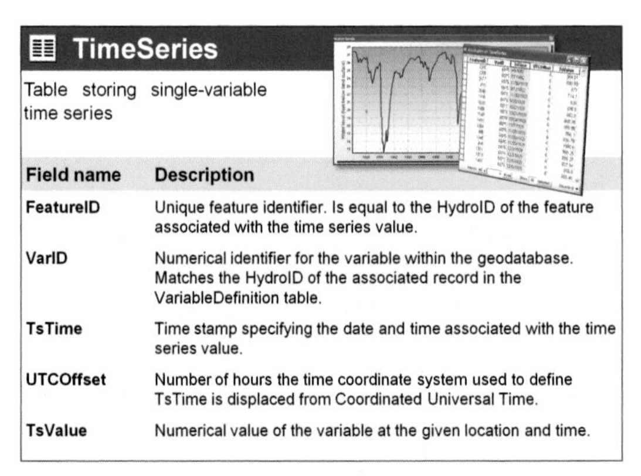

TimeSeries

Table storing single-variable time series

Field name	Description
FeatureID	Unique feature identifier. Is equal to the HydroID of the feature associated with the time series value.
VarID	Numerical identifier for the variable within the geodatabase. Matches the HydroID of the associated record in the VariableDefinition table.
TsTime	Time stamp specifying the date and time associated with the time series value.
UTCOffset	Number of hours the time coordinate system used to define TsTime is displaced from Coordinated Universal Time.
TsValue	Numerical value of the variable at the given location and time.

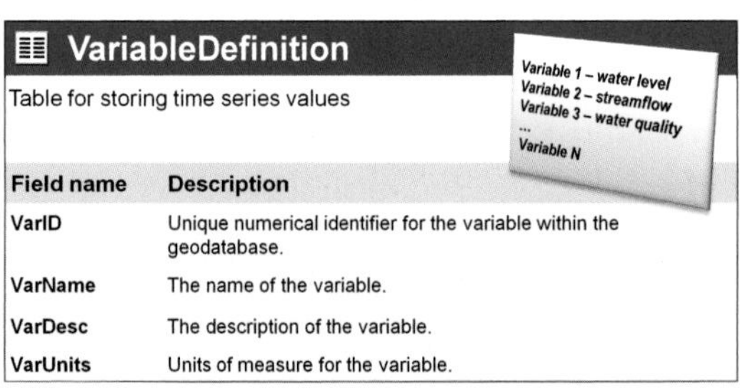

VariableDefinition

Table for storing time series values

Field name	Description
VarID	Unique numerical identifier for the variable within the geodatabase.
VarName	The name of the variable.
VarDesc	The description of the variable.
VarUnits	Units of measure for the variable.

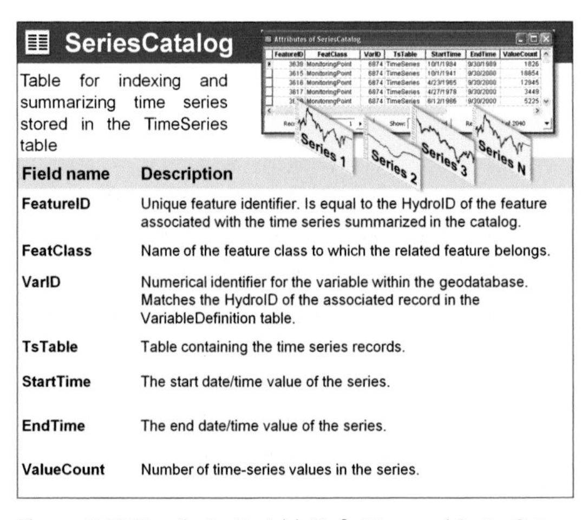

SeriesCatalog

Table for indexing and summarizing time series stored in the TimeSeries table

Field name	Description
FeatureID	Unique feature identifier. Is equal to the HydroID of the feature associated with the time series summarized in the catalog.
FeatClass	Name of the feature class to which the related feature belongs.
VarID	Numerical identifier for the variable within the geodatabase. Matches the HydroID of the associated record in the VariableDefinition table.
TsTable	Table containing the time series records.
StartTime	The start date/time value of the series.
EndTime	The end date/time value of the series.
ValueCount	Number of time-series values in the series.

Figure 2.12 TimeSeries,VariableDefinition, and SeriesCatalog tables for storing temporal data.

complete description of these tables is presented in chapter 7.

The Arc Hydro framework is simple and is easily created, yet it is powerful enough to support a range of applications. First, the relationships in the framework support a variety of querying and visualization capabilities. Some simple example queries are: "show me all the domestic wells in the Edwards Aquifer," or "select all streams and water bodies within the Guadalupe River basin," or "select all stream segments related with the Edwards Aquifer outcrop." More sophisticated queries joining features and TimeSeries (e.g., using the Well-TimeSeries relationship) can be used to plot observations such as streamflow and groundwater levels over time, and we can design special SQL-based queries to generate maps of groundwater levels and groundwater quality for mapped aquifers (see chapter 7 for more details on creating spatial-temporal views of groundwater observations). Because the framework represents both surface water and groundwater features, we can visualize and analyze data for surface water and groundwater in conjunction.

Implementation

This section outlines the main steps involved in creating and populating the Arc Hydro framework. The first step is to create the geodatabase classes necessary for your project. You can use the XML schema as a template to create the framework classes, or you can create them with ArcCatalog. After creating the classes, you set the spatial reference appropriate for your project, add project specific attributes, add relationships, create additional tables and feature classes if necessary, and add coded value domains for selected attributes. The next step is to document your newly created data model.

Once the data model is defined you can import data into your geodatabase. Next, HydroIDs are assigned to the features, and relationships are established. New features can be created and attributes calculated. Finally, you can use the results of this process to create products such as maps, scenes, and reports. The following checklist provides a summary of the main steps for creating the Arc Hydro framework.

Checklist

1. Create the classes of the Arc Hydro framework (manually using ArcCatalog or by importing from an XML schema)
2. Define the spatial reference, add project specific classes, attributes, relationships, and domains as necessary
3. Document the datasets and changes made to the data model
4. Import data into the framework classes (e.g., streams, wells, aquifers, time series)
5. Assign key attributes to uniquely identify the features and establish relationships
6. Apply tools to create new features and calculate attributes
7. Visualize data and create products (maps, scenes, reports)

The following example of an Arc Hydro Framework representation of the Guadalupe Basin and the Edwards Aquifer in Texas illustrates the use of the framework to represent surface water and groundwater features. After creating the feature classes of the framework, the first step is to import data into the framework geodatabase. In the following example we added streams, water bodies, and watersheds from the National Hydrography Dataset (NHD). HydroIDs were assigned to all features so they could be uniquely identified within the geodatabase. HydroCodes were established to maintain the original public identifier of the features. For example, river reaches from NHD were imported into the WaterLine feature class (figure 2.13). Each line was assigned a HydroID that became its unique identifier within the Arc Hydro geodatabase. In addition, the NHD reach codes were imported to the HydroCode attribute of the of the line features. For example, the river reach representing the Blanco River was indexed with a HydroID of 8753, which became its unique identifier within the geodatabase. The NHD code for this reach is 12100203000078, thus the HydroCode given to this station is 12100203000078 to represent the original reach code. You can have multiple NHD river segments with the same reach identifier, thus you may have multiple WaterLine features each with a unique HydroID sharing the same HydroCode.

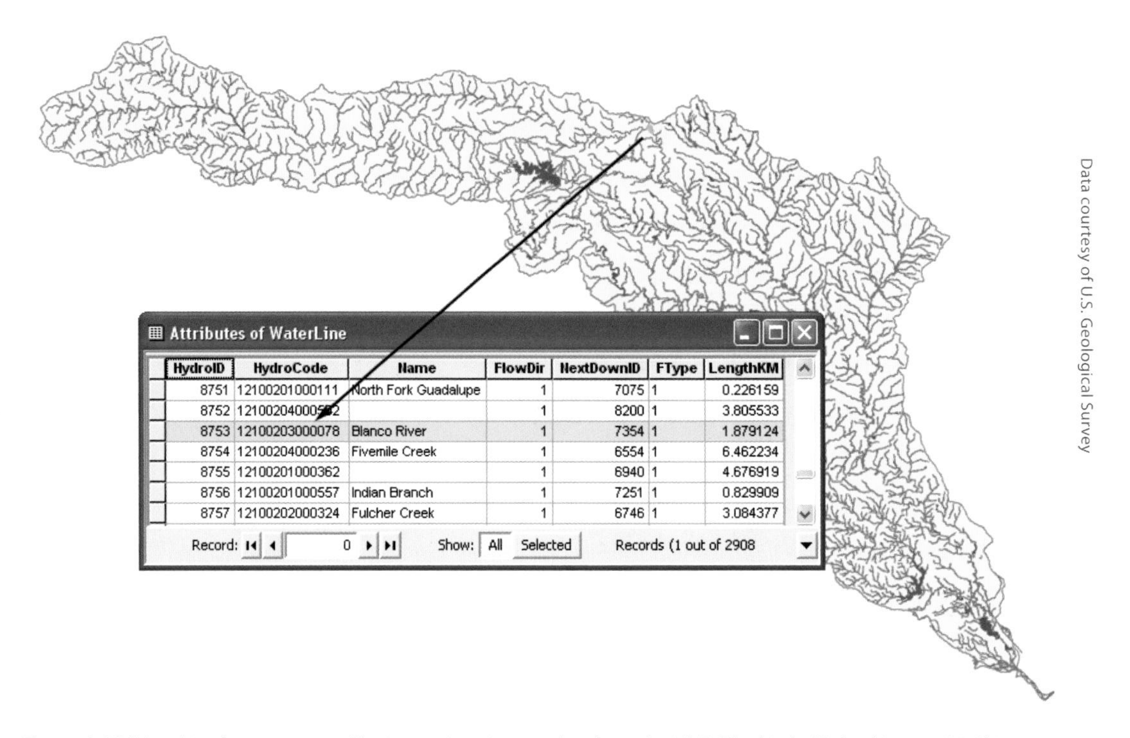

Figure 2.13 WaterLine features created by importing river reaches from the NHD. The HydroID (in this case 8753) is a unique feature identifier within the geodatabase, and HydroCode (12100203000078) is the public identifier that relates to external information systems, in this case NHD.

Aquifer boundaries and well features from the Texas Water Development Board database were imported into the Aquifer and Well feature classes. Similarly to the NHD features, aquifers and wells were also assigned HydroID and HydroCode values to uniquely identify them within the geodatabase and to maintain a connection to the original information systems. Additional attributes such as FType were also assigned to distinguish between types of wells and zones within the aquifer. For example, figure 2.14 shows aquifer features symbolized by the FType attribute to distinguish between outcrop and confined zones of the aquifer. The selected feature in the following map (HydroID = 114) shows an unconfined area of the Edwards Aquifer.

Streams that flow across the unconfined zones of the Edwards Aquifer are possible recharge

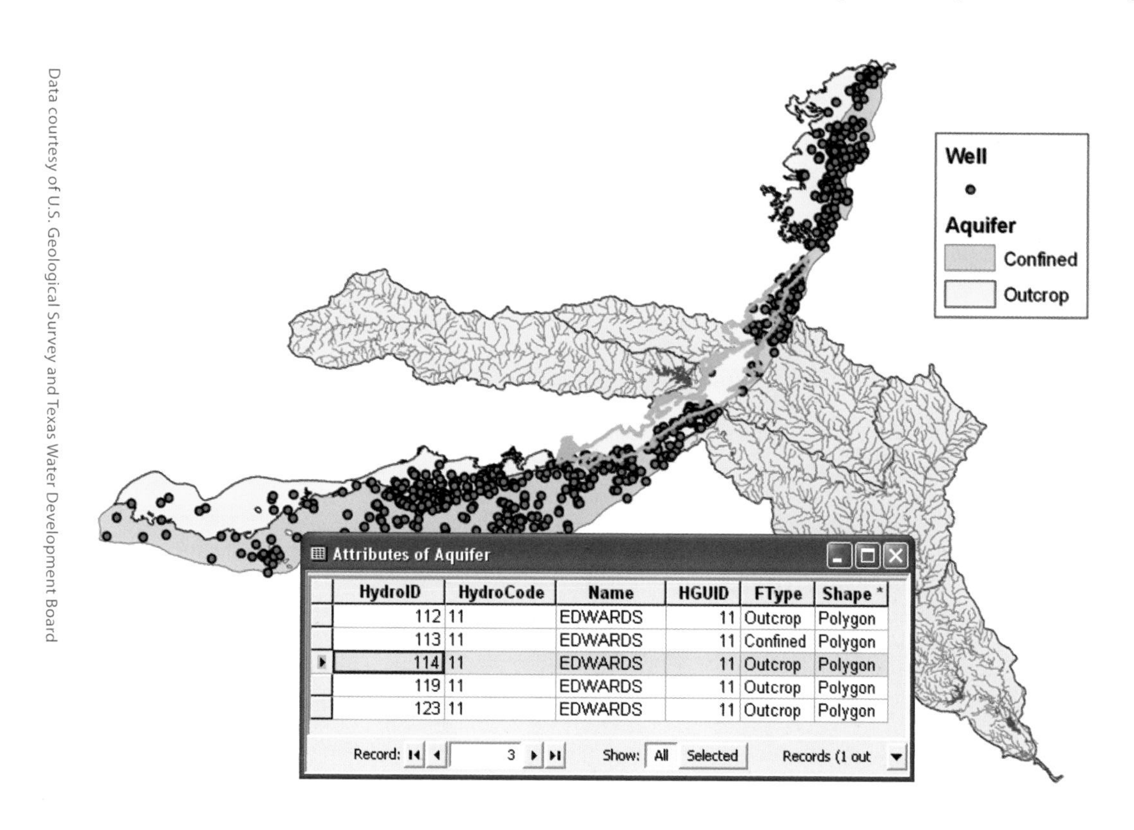

Data courtesy of U.S. Geological Survey and Texas Water Development Board

Figure 2.14 Example of Aquifer and Well features.

features to the aquifer. Thus, an association can be established between river reaches flowing over the unconfined zones and the aquifer below. This is done by adding an AquiferID attribute to the WaterLine feature class and setting it equal to the HydroID of the Aquifer feature to which WaterLine features are related. In this case, the river reaches that intersect the Edwards Aquifer outcrop are assigned with an AquiferID of 114, which is equal to the HydroID of the feature representing the outcrop zone of the Edwards Aquifer (figure 2.15).

Monitoring stations representing stream gages along the river network from the USGS National Water Information System (NWIS) were imported into the MonitoringPoint feature class. HydroID and HydroCode attributes were assigned to the stations. For example, a USGS monitoring station at New Braunfels, Texas, records the streamflow coming out of Comal Springs. The HydroID given

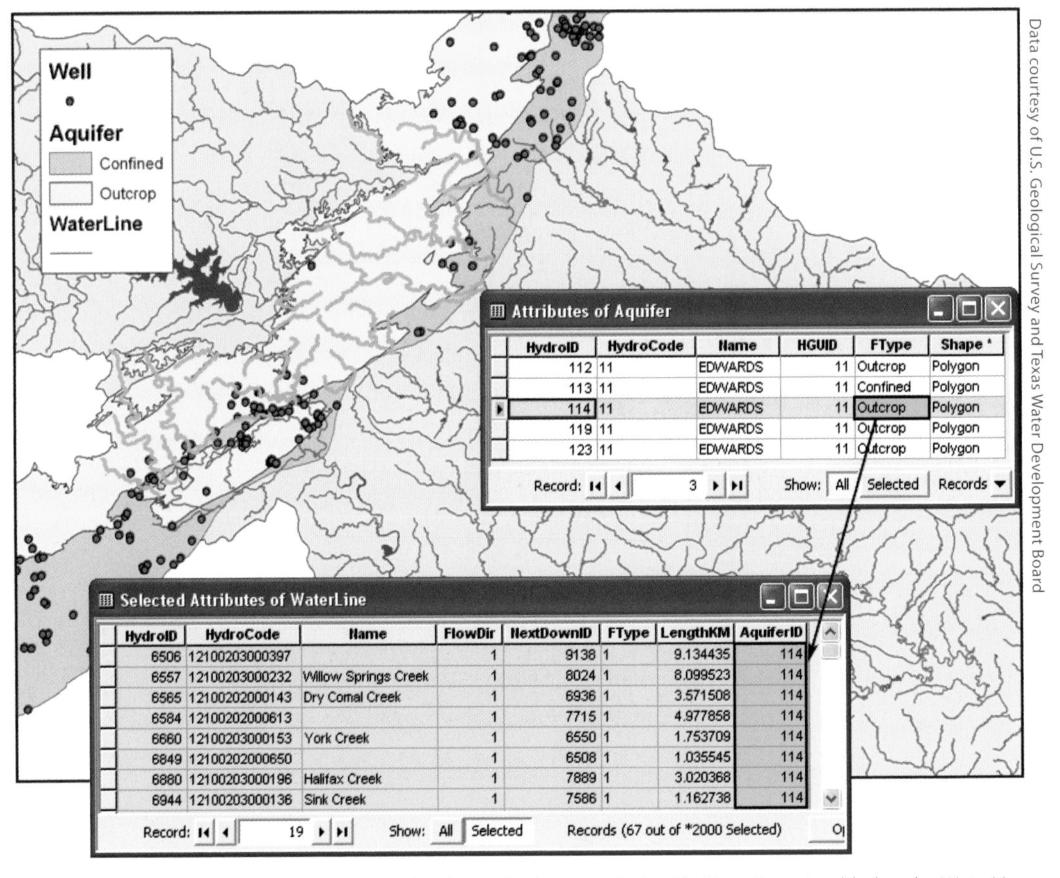

HydroID	HydroCode	Name	HGUID	FType	Shape*
112	11	EDWARDS	11	Outcrop	Polygon
113	11	EDWARDS	11	Confined	Polygon
114	11	EDWARDS	11	Outcrop	Polygon
119	11	EDWARDS	11	Outcrop	Polygon
123	11	EDWARDS	11	Outcrop	Polygon

HydroID	HydroCode	Name	FlowDir	NextDownID	FType	LengthKM	AquiferID
6506	12100203000397		1	9138	1	9.134435	114
6557	12100203000232	Willow Springs Creek	1	8024	1	8.099523	114
6565	12100202000143	Dry Comal Creek	1	6936	1	3.571508	114
6584	12100202000613		1	7715	1	4.977858	114
6660	12100203000153	York Creek	1	6550	1	1.753709	114
6849	12100202000650		1	6508	1	1.035545	114
6880	12100203000196	Halifax Creek	1	7889	1	3.020368	114
6944	12100203000136	Sink Creek	1	7586	1	1.162738	114

Data courtesy of U.S. Geological Survey and Texas Water Development Board

Figure 2.15 WaterLine features can be associated with Aquifer features. An AquiferID attribute is added to the WaterLine feature class and is set equal to the HydroID of an Aquifer feature.

to this feature is 2894, and the USGS code for this station is 08168710, thus the HydroCode for the station was set as "08168710" to represent the original station number.

In addition to spatial features, time series of streamflow and water levels from the NWIS and the TWDB groundwater database, respectively, were imported into the TimeSeries table. Each time-series record was indexed with the HydroID of the feature to which it relates. Thus, each time-series record in the table can be located in space through the association with its related feature. For example, streamflow records recorded at the streamflow gage at Comal Springs (figure 2.16) were indexed with the HydroID of the MontoringPoint feature (2894). Similarly, time series

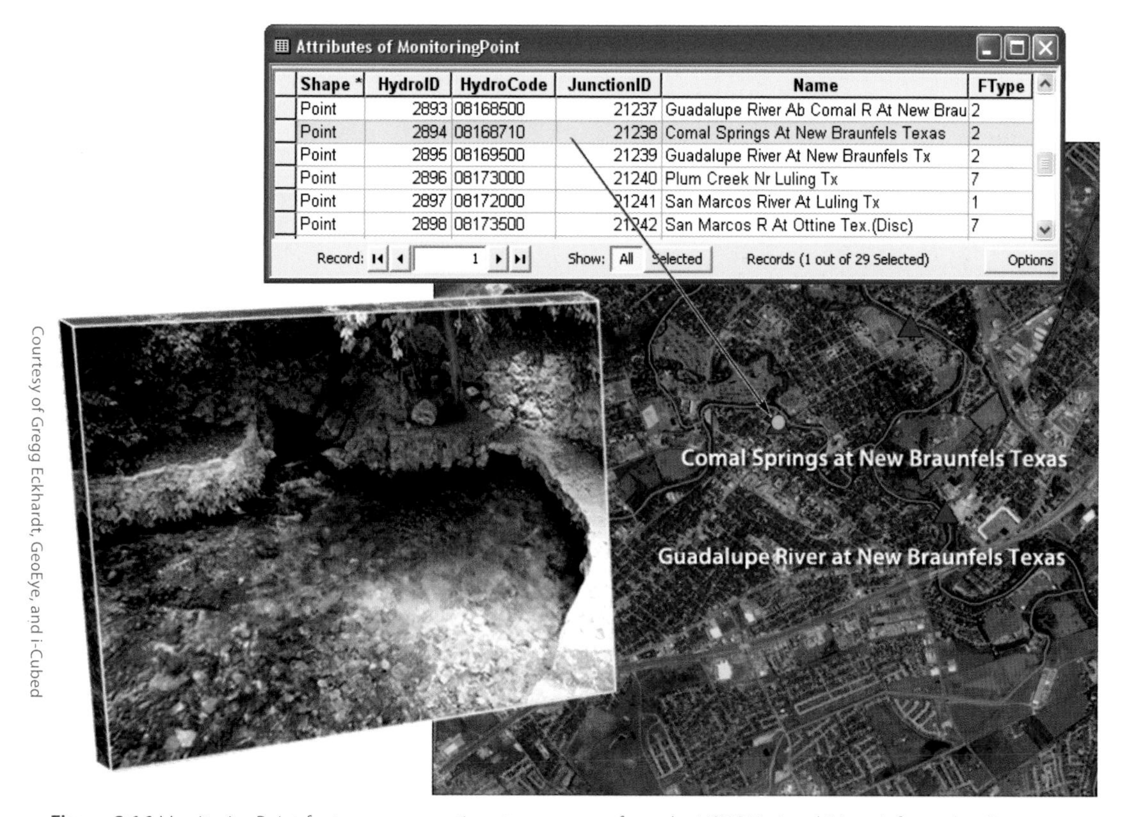

Courtesy of Gregg Eckhardt, GeoEye, and i-Cubed

Figure 2.16 MonitoringPoint features representing stream gages from the USGS National Water Information System.

representing water levels were imported and indexed so they relate to Well features. The example below (figure 2.17) shows two time-series plots: the red plot represents groundwater levels (feet below land surface) at a well in the Edwards Aquifer adjacent to Comal Springs in New Braunfels Texas, and the blue plot shows streamflow at a gaging station downstream from the springs. The plots demonstrate the strong connectivity between streamflow and groundwater levels in the Edwards Aquifer. During the 1950s drought groundwater levels dropped sharply, resulting in Comal Springs going dry for a few months.

References

Arctur, David, and Michael Zeiler. 2004. *Designing Geodatabases: Case Studies in GIS Data Modeling*. Redlands, California: Esri Press.

Booch, Grady, James Rumbaugh, and Ivar Jacobson. 1999. *The unified modeling language user guide*. Addison-Wesley Professional.

Zeiler, Michael. 1999. *Modeling Our World: The Esri Guide to Geodatabase Design*. Redlands, California: Esri Press.

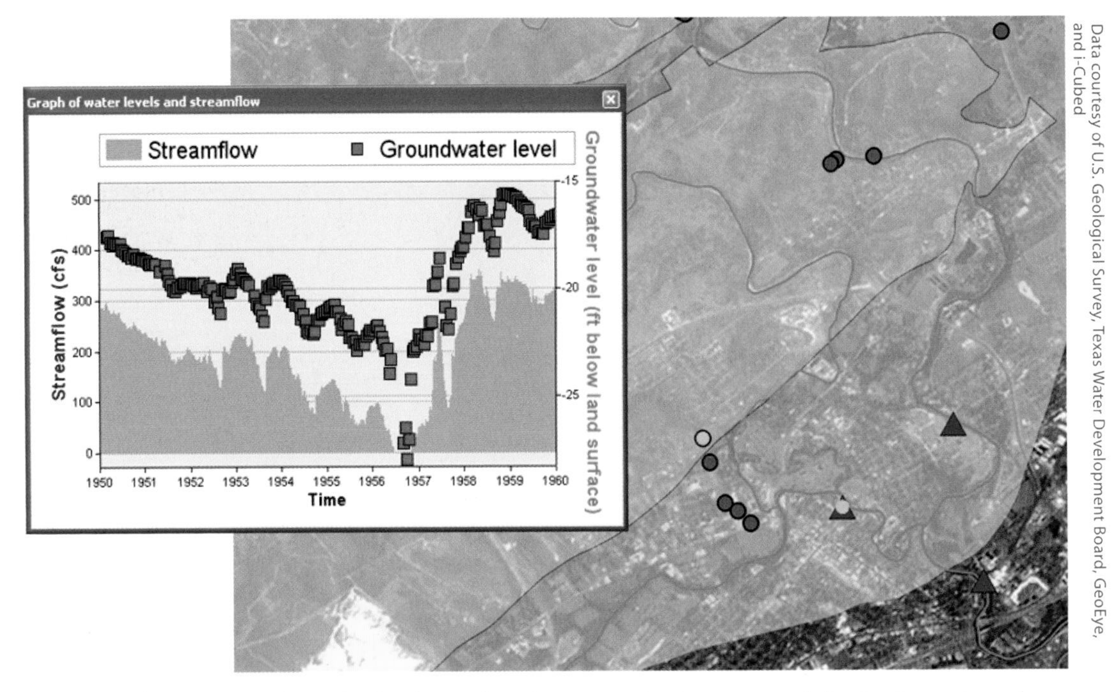

Data courtesy of U.S. Geological Survey, Texas Water Development Board, GeoEye, and i-Cubed

Figure 2.17 The time series of water level and streamflow demonstrate the strong connectivity of surface water and groundwater in the Edwards Aquifer. Groundwater levels shown in the plot are from the Texas Water Development Board groundwater database, and streamflow are from the USGS NWIS system.

Three-dimensional ArcGIS for subsurface representation

GIL STRASSBERG

THE ABILITY TO STORE, VISUALIZE, QUERY, AND ANALYZE 3D PHENOMENA IS A well-established field in computer science. In the past decade, much advancement has been made in the field of computer-aided visualization as software, hardware, and viewing devices have been improved to support better design, science, and entertainment. This evolution has not bypassed the field of geographic information, and GIS now offers tools for visualizing multidimensional phenomena, including animation of 3D and time-varying datasets. Three-dimensional GIS is relevant to many industries and disciplines. GIS users in fields ranging from petroleum and groundwater to atmospheric science and oceanography are finding similarities in the way they view and store 3D datasets. This chapter starts with a review of 3D GIS for subsurface characterization followed by a description of 3D geodatabase datasets and how these are used for subsurface representation.

3D GIS for subsurface characterization

Groundwater movement within the subsurface is truly a 3D phenomenon. Water can enter the subsurface as diffuse recharge over an outcrop area or as focused recharge from streams and water bodies, then travel horizontally within an aquifer unit down to an underlying aquifer, or down and then up to discharge at a spring or into the ocean. The description of groundwater movement within the subsurface also requires representations of hydrogeologic formations, flow barriers and conduits, and the heterogeneity of strata. For the geology, mining, and petroleum industries, specialized software packages have been developed to create sophisticated 3D models of the subsurface. This branch of GIS is also known as geoscientific information systems (GSIS) or geomodeling systems. These differ from traditional GIS in their capabilities to represent complex 3D objects using either surface representations or volume objects, mainly voxels, which are equivalent to a 3D raster (Turner 2000). GSIS also has the capabilities to rapidly display, edit, and interpolate 3D objects to support subsurface conceptualization. GSIS is based on "geo-objects" which are distinctive subsurface features with measurable spatial boundaries in 3D (Fisher 1993). Geo-objects describe discrete entities such as a rock layer, fault, or a volume element. A combination of geo-objects forms a geomodel (figure 3.1) that provides an abstract digital representation of a part of the earth's subsurface (Apel 2006).

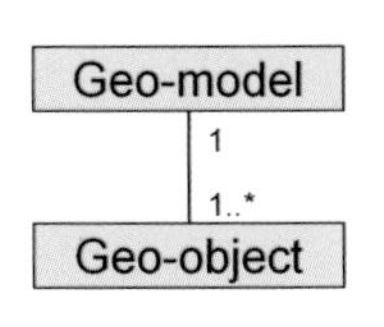

Figure 3.1 A geo-model is an abstract representation of a part of the Earth's subsurface. Geomodels are composed of one or more geo-objects that are distinctive subsurface features.

Geomodels defining hydrogeologic units are created to understand the movement of groundwater within the subsurface. Groundwater software applications use geologic maps, borehole stratigraphy, cross-sections, and terrain data to construct 3D models of the subsurface, and these form the base for developing numerical flow and transport simulation models. An example of such an application is the groundwater modeling system (GMS) used to create 3D hydrogeologic models and develop groundwater simulation models. Designing a groundwater data model requires us to answer two important questions: How well can we represent geo-objects within the geodatabase? How can we best use current ArcGIS datasets and tools to create, edit, and display 3D geo-objects? To answer these questions, we must first understand the workflow by which geo-objects are defined; how data is collected, stored, and analyzed; and what data structures are necessary for creating a geomodel representing a groundwater system.

When conceptualizing a 3D groundwater system we usually start by abstracting the subsurface into hydrogeologic units. A hydrogeologic unit is a rock unit or zone that because of its hydraulic properties has a distinct influence on groundwater flow and storage. Hydrogeologic units include both aquifer units and confining units (also known as aquitards). These units have horizontal and vertical extents and are considered 3D.

One of the major constraints for establishing an accurate representation of the subsurface is the lack of field data. While remotely sensed datasets are advancing our understanding and ability to model the movement of water on the earth's surface and in the atmosphere, datasets describing the subsurface are relatively scarce. Classification of earth strata is based on samples extracted from boreholes (such as cores and cuttings), or inferred from geophysical logs. The borehole data are usually stored as a set of hydrogeologic contacts, with each contact representing the top of a hydrogeologic unit. The high cost of drilling prohibits collection of dense datasets. Thus, estimating the extent of hydrogeologic units is usually based on upscaling point borehole data to a larger area through interpolation procedures. The borehole point observations and the interpolated results are managed and visualized with a number of common data structures including 3D points and lines representing borehole contacts and intervals, cross-sections and fence diagrams, surfaces, and volumes (figure 3.2).

Hydrostratigraphy along a borehole can be described as a set of 3D points that represent hydrogeologic contacts or as 3D lines representing the hydrogeologic units as intervals along the borehole. Cross-sections connecting the borehole data are created for display purposes by interpolating between known data points projected on a vertical plane. The cross-section itself can be viewed in 2D. By editing the cross-section we can add imaginary data points between observations along the cross-section's plane. Fence diagrams are collections of cross-sections viewed in 3D. Borehole data and added points from cross-sections (together with terrain data and geologic maps) are the basis for creating surfaces representing the top and bottom boundaries of hydrogeologic units. Using borehole points we can interpolate surfaces representing the top and bottom boundaries of hydrogeologic units. By repeating this process we generate a set of "stacked" surfaces defining the vertical extent of hydrogeologic units over a defined area. We then use the surfaces to generate volume objects that represent the hydrogeologic units as individual entities. Volume objects capture the "true" (in this sense we regard our conceptualization of the system as the truth) geometry of hydrogeologic units because they describe the horizontal and vertical extent of the unit. Additional cross-sections can be created by "slicing" through volume objects to view the interior of the solids along planes of interest.

The above process highlights the major geo-objects necessary for describing a geomodel. These include 3D objects for representing points and lines along boreholes, cross-sections, surfaces, and volumes. The following sections illustrate how geodatabase datasets represent such objects.

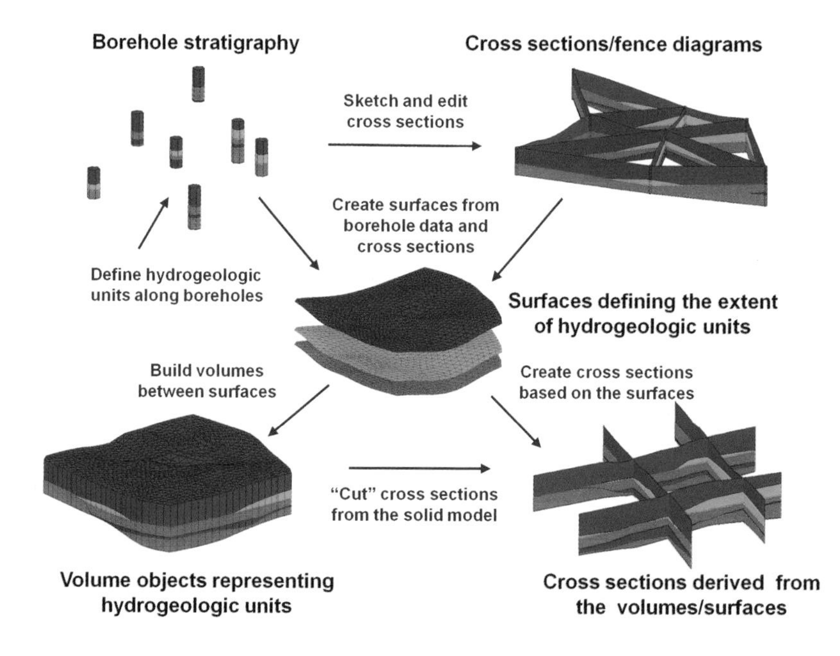

Figure 3.2 Process of creating a 3D description of hydrogeologic units. Blue text represents the processes and black text shows the derived data products.

Z-enabled features in the geodatabase

Feature classes within the geodatabase can store 2D and 3D geometries. When creating a new feature class, the default is a 2D representation of features with longitude (x) and latitude (y) coordinates, but we can enable features to store coordinates in the vertical (z) dimension and any z-enabled feature class can hold 3D geometries (figure 3.3).

Once a feature class is defined as z-enabled, we store 3D features in it by assigning vertical coordinates to the vertices of the geometry. In the geodatabase, z-enabled features are distinguished by adding a "Z" to the geometry name (figure 3.4). For example, a Point feature becomes PointZ, a Line feature LineZ, and a Polygon feature becomes PolygonZ. The multipatch is a unique geometry type as it is always z-enabled.

To create a 3D feature we need to edit the z coordinate of each of the feature's vertices. This can be done in a regular edit session within ArcMap or programmatically using custom tools. Take for example the polygon shown in figure 3.5; because it is z-enabled, each vertex is defined by coordinates in x, y, and z which can be viewed and modified in an edit session. This allows us to construct a 3D polygon representing an area in 3D space. If we view the polygon feature in Arc Map, we see it as 2D, but when we load it into ArcScene, the feature displays in 3D. A similar approach can be used to create 3D points and lines. Keep in mind that these 3D features are not truly 3D objects, and that 3D operations and calculations cannot be performed directly with the z-enabled

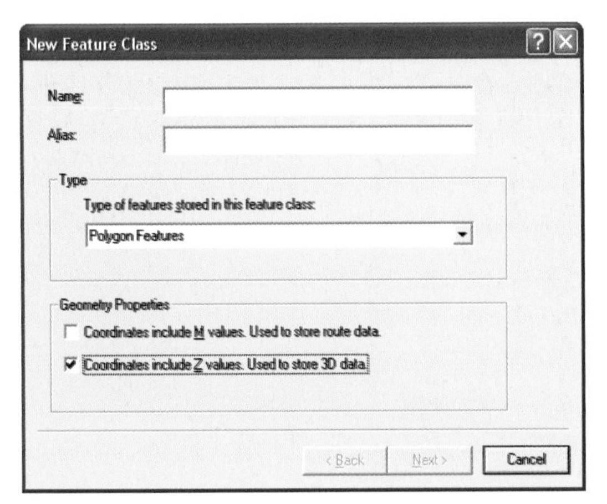

Figure 3.3 Z-enabling a feature class enables the storage of 3D geometries (points, lines, polygons, and multipatches) within the geodatabase. When creating a new feature class in ArcCatalog, we can select to include Z values, thus enabling the storage of vertical coordinates in the geometry.

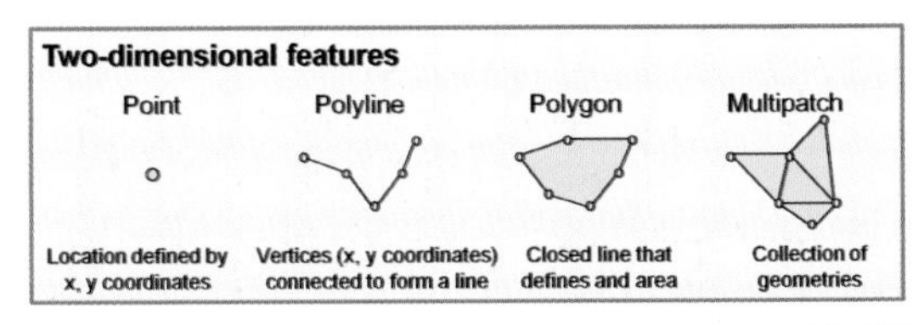

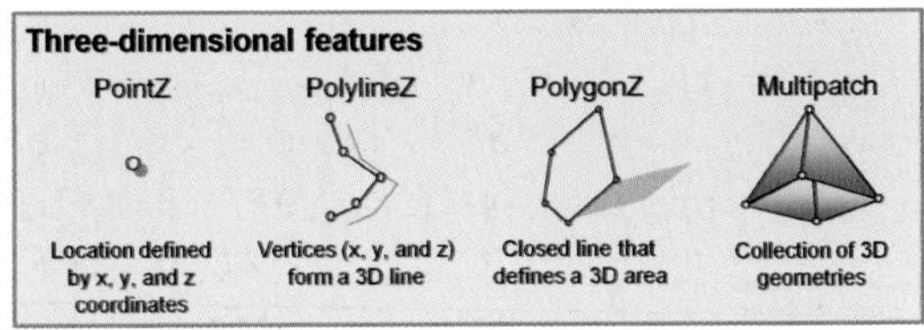

Figure 3.4 Features that are z-enabled support the definition of a vertical coordinate and enable the representation of 3D geometries.

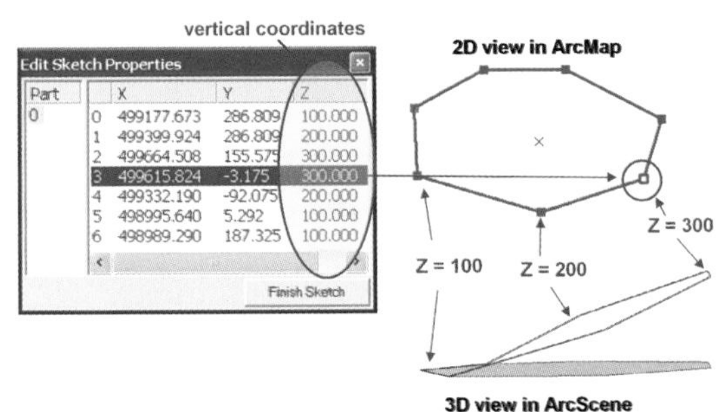

Figure 3.5 Three-dimensional coordinates of a z-enabled polygon feature. Each vertex in the geometry is assigned a z coordinate, and the feature displays in 3D context when viewed in ArcScene.

features. For example, a z-enabled polygon feature does not automatically calculate its 3D area, and a z-enabled line feature does not show its true 3D length. Rather, the features show the 2D projected area and length as they are viewed in ArcMap. Also, operations such as intersection or buffering which are commonly applied on 2D features cannot be applied in 3D. Thus, z-enabled objects are somewhat limited but still provide a data structure to store and display 3D geometries within the ArcGIS environment.

Creating volume (multipatch) objects

A multipatch geometry is composed of 3D rings and triangles and represents objects that occupy a 3D area or volume. These can be geometric objects such as a cube or a sphere, or represent real-world objects such as buildings or trees (`http://support.esri.com`). In the groundwater data model, multipatches represent volume objects usually defined as a set of triangular elements defining the surface boundary of the volume feature. The interior of the volume is not discretized, and the attributes of the feature apply to the volume defined by the multipatch (similar to the area of a polygon in 2D space). Several methods are available for creating multipatch geometries. The simplest method is to create a 3D multipatch by extruding a 2D polygon vertically (either up or down). The extrusion process takes the area defined by the polygon feature and creates a 3D multipatch feature that represents a 3D volume (figure 3.6). The extrusion process starts with a base geometry that itself can be z-enabled and can have a different z value at each vertex. The base geometry is then extruded either by adding a constant height to the base geometry or by extruding the base geometry to a defined height.

Another option for creating multipatch features is to extrude a base geometry between two surfaces. This method allows us to represent a volume of space between two surfaces over a given area. For example, we can take two surfaces defining the top and bottom of a hydrogeologic unit and apply the "Extrude Between" geoprocessing tool (part of the 3D Analyst extension) to create a complex multipatch feature representing the volume between the two surfaces (figure 3.7). By repeating this process over a set of "stacked" surfaces representing a sequence of hydrogeologic units, a set of volumes can be defined, forming a 3D representation of a hydrogeologic model. This method is implemented in the Arc Hydro Groundwater tools to create a set of GeoVolumes from raster surfaces indexed in a raster catalog (see chapter 6 for a more detailed description of this process).

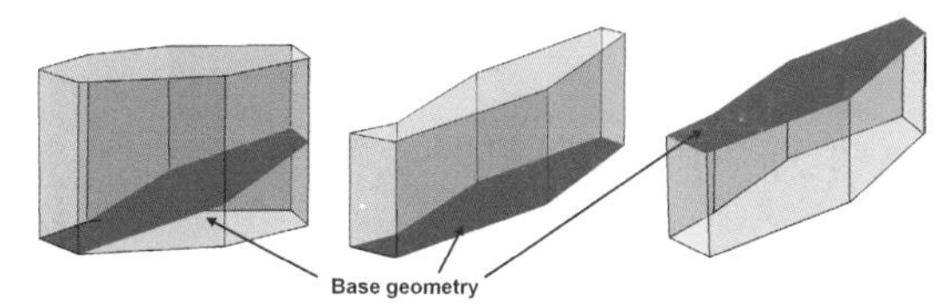

Figure 3.6. Multipatch features created by extruding a z-enabled base geometry (shown in blue): at left, the base geometry is extruded between two constant heights; in the middle, the base geometry is extruded upward by adding a constant height to its base heights; at right, the base geometry is extruded downward.

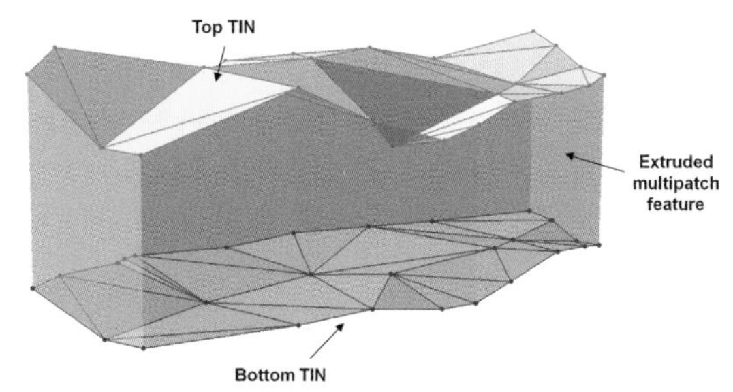

Figure 3.7 A volume object (multipatch) created by extruding a polygon between two TIN surfaces using the Extrude Between geoprocessing tool.

More complex 3D volumes can be created by developing a surface representation of the volume, also known as a polyhedron. A polyhedron is a region of space whose boundary is composed of a finite number of flat polygonal faces (O'Rourke 1998). Polyhedron surfaces are commonly expressed as surface triangulations to simplify the data structure and construction of the geometry, and many software applications use triangulated surfaces to represent 3D volume objects. While ArcGIS currently does not include algorithms and editing tools to construct complex volumes, these objects can still be stored as multipatch features and viewed within ArcScene. In this manner, volume objects created in specialized software applications can be imported into the geodatabase and viewed in context with other GIS datasets (one of the Arc Hydro Groundwater tools can import volumes from XML documents into a multipatch feature class). For example, figure 3.8 shows a text file describing a set of volume objects. Each object is described by a set of attributes such as its identifier, name, and the number of vertices and triangles from which it is composed. The geometry of the object is constructed as a collection of triangles defined by a set of 3D vertices.

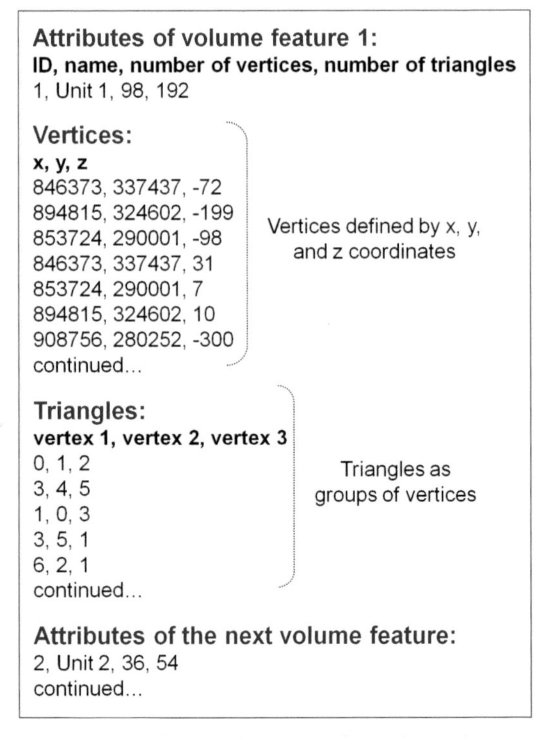
Attributes of volume feature 1:
ID, name, number of vertices, number of triangles
1, Unit 1, 98, 192

Vertices:
x, y, z
846373, 337437, -72
894815, 324602, -199
853724, 290001, -98
846373, 337437, 31
853724, 290001, 7
894815, 324602, 10
908756, 280252, -300
continued...

Triangles:
vertex 1, vertex 2, vertex 3
0, 1, 2
3, 4, 5
1, 0, 3
3, 5, 1
6, 2, 1
continued...

Attributes of the next volume feature:
2, Unit 2, 36, 54
continued...

Figure 3.8 Text file describing a set of 3D volume objects. Each volume contains a set of attributes (e.g., ID and Name), and the 3D geometry of the volume is defined by a set of triangles composed of 3D vertices.

This volume object is stored in the geodatabase as a multipatch feature constructed by reading (programmatically) the text file and building the 3D geometry. Properties of the volume (ID, Name) are stored as attributes of the feature and can be accessed through the feature class's attribute table (figure 3.9).

References

Apel, M. 2006. From 3D geomodelling systems towards 3D geosciences information systems: Data model, query functionality, and data management. *Computers & Geosciences* 32: 222–229.

Fisher, T. R. 1993. Use of 3D Geographic Information Systems in Hazardous Waste Site Investigations. Pages xxiii, 488, [488] of col. plates in *Environmental modeling with GIS*, ed. M. F. Goodchild, B. O. Parks, and L. T. Steyaert, Oxford University Press, New York.

O'Rourke, J. 1998. Computational geometry in C, 2nd edition. Cambridge University Press, Cambridge, UK, ; New York, New York.

Turner, A. K. 2000. Geoscientific Modeling: Past, Present, and Future in *Geographic information systems in petroleum exploration and development: AAPG computer applications in geology no. 4*, ed. T. C. Coburn and J. M. Yarus, American Association of Petroleum Geologists, Tulsa, Oklahoma.

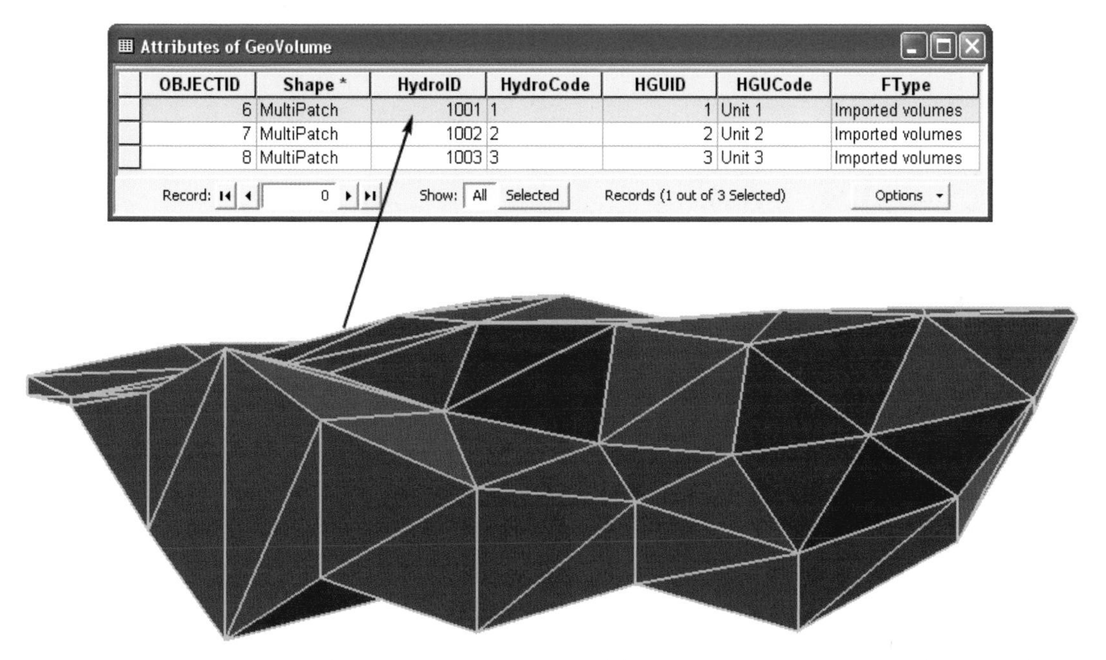

OBJECTID	Shape *	HydroID	HydroCode	HGUID	HGUCode	FType
6	MultiPatch	1001	1	1	Unit 1	Imported volumes
7	MultiPatch	1002	2	2	Unit 2	Imported volumes
8	MultiPatch	1003	3	3	Unit 3	Imported volumes

Figure 3.9 A volume object stored in a geodatabase as a multipatch feature. The volume is constructed from a set of triangular elements.

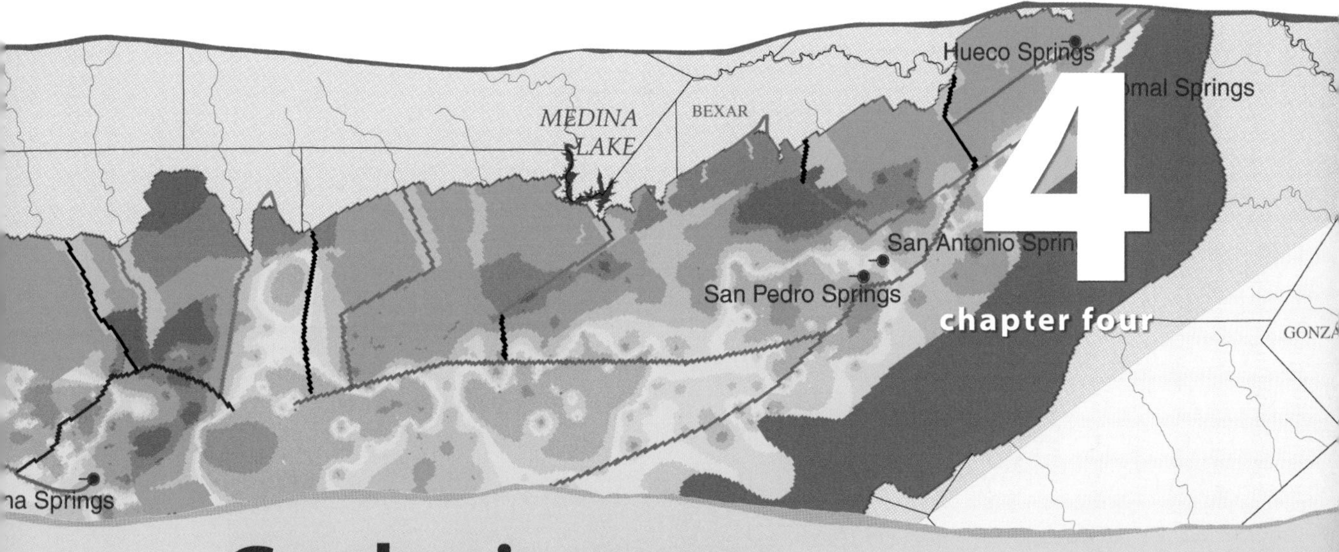

4 chapter four

Geologic maps

GIL STRASSBERG

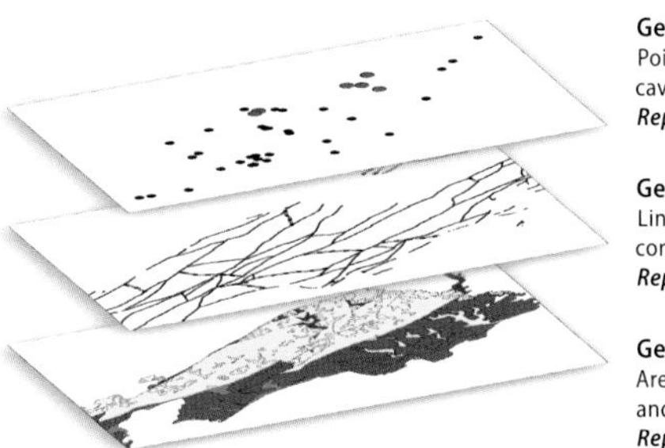

GeologyPoint
Point features from geologic maps such as springs, caves, sinks, and observation points.
Representation: Point features.

GeologyLine
Line features from geologic maps such as faults, contacts, and dikes.
Representation: Polyline features.

GeologyArea
Areal features from geologic maps such as rock units and alteration zones.
Representation: Polygon features.

TO DEVELOP A HYDROGEOLOGIC MODEL ONE MUST FIRST UNDERSTAND THE GEOLOGY in the area of interest. Geologic maps are the most common data products used to describe the geology of a region. Geologic maps are developed on a wide range of scales, from continental maps to site scale maps. For groundwater purposes, geologic maps provide important information about the geologic formations in the area and the locations of geologic features that may serve as water conduits or barriers. Geologic maps are complex because they describe the 3D distribution of material within the subsurface. For the Arc Hydro Groundwater data model, a simplified set of features is used to represent data from geologic maps. These features provide a basic placeholder to include spatial data from geologic maps in groundwater projects.

Geologic maps and geologic map databases

Geologic maps are cartographic products containing information about the kinds of earth materials in a specific geographic area, the boundaries that separate them, and the geologic structures that have deformed them (Geologic Data Subcommittee, Federal Geographic Data Committee 2006). These maps represent 3D geologic features in a 2D mapping environment. To do this, the maps usually contain special graphical elements such as cross-sections, graphic legends, and text blocks explaining the geology of the region portrayed. In many cases, geologic maps are the basis for mapping groundwater features such as aquifer units. In turn, aquifer boundaries described in aquifer maps are closely related to geologic formations defined in geologic maps. For example, figure 4.1 shows a generalized geologic map and a map of the major aquifers of the United States. The maps demonstrate the close spatial relationships between geologic formations and aquifers.

Standards to support the creation of geologic maps have been established for many years by different agencies. Guides for creating geologic maps were developed soon after the USGS was established in 1879 (Geologic Data Subcommittee, Federal Geographic Data Committee 2006). In recent years, the adoption of GIS to collect geologic data in the field, archive the data, and produce maps has led to the design of standards for representing digital geologic maps. These standards define a geologic map database, which is a digitally compiled collection of spatial (geographically referenced) and descriptive geologic information about a specific geographic area (Geologic Data Subcommittee, Federal Geographic Data Committee 2006). One example is the North American Geologic Map Data Model (NADM) that aims to develop standardized methodologies for the storage, manipulation, analysis, management, and distribution of digital geologic-map information (`http://www.nadm-geo.org`). Another example is the National Geologic Map Database (NGMDB), a collaborative effort to create a single national standard for the digital cartographic representation of geologic map features, primarily involving the USGS and the Association of American State Geologists (`http://ngmdb.usgs.gov`).

Geologic map databases contain information on the geographic location, geometry, and attributes of geologic features. The fundamental data

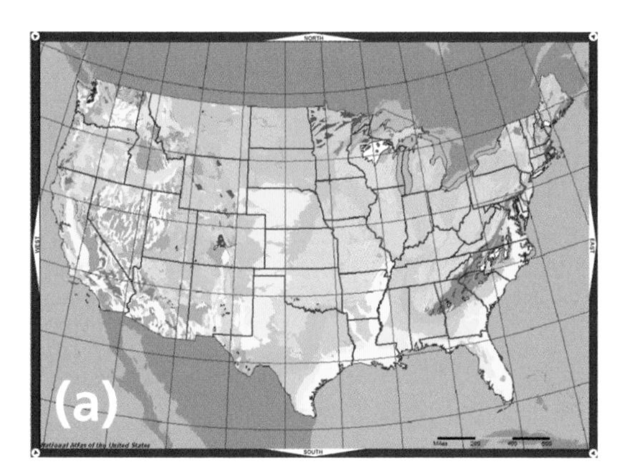

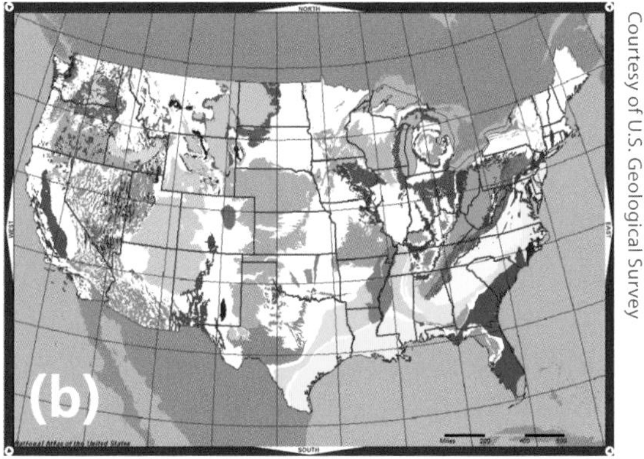

Courtesy of U.S. Geological Survey

Figure 4.1 a) Generalized geologic map of the United States, and b) Principal aquifers of the United States. Map units are colored to represent different geologic units and aquifers.

elements of a geologic map database are lines (e.g., contacts and faults), points (e.g., bedding attitudes and sample locations), and polygons (e.g., map-unit areas and zones of alteration). Geologic maps are generated based on the objects in the geologic map database by symbolizing features and labeling the features with the appropriate attributes stored in the database. An example of a generic geologic map database implemented within ArcGIS is the Arc Geology geodatabase design (Raines et al. 2007). This design takes a simplistic approach for representing geologic features, cross-sections, and legend graphics, using sets of point, line, and polygon feature classes with a set of common feature attributes. Arc Geology enables the production of geologic maps from features stored within the geodatabase by simple built-in ArcGIS tools (figure 4.2).

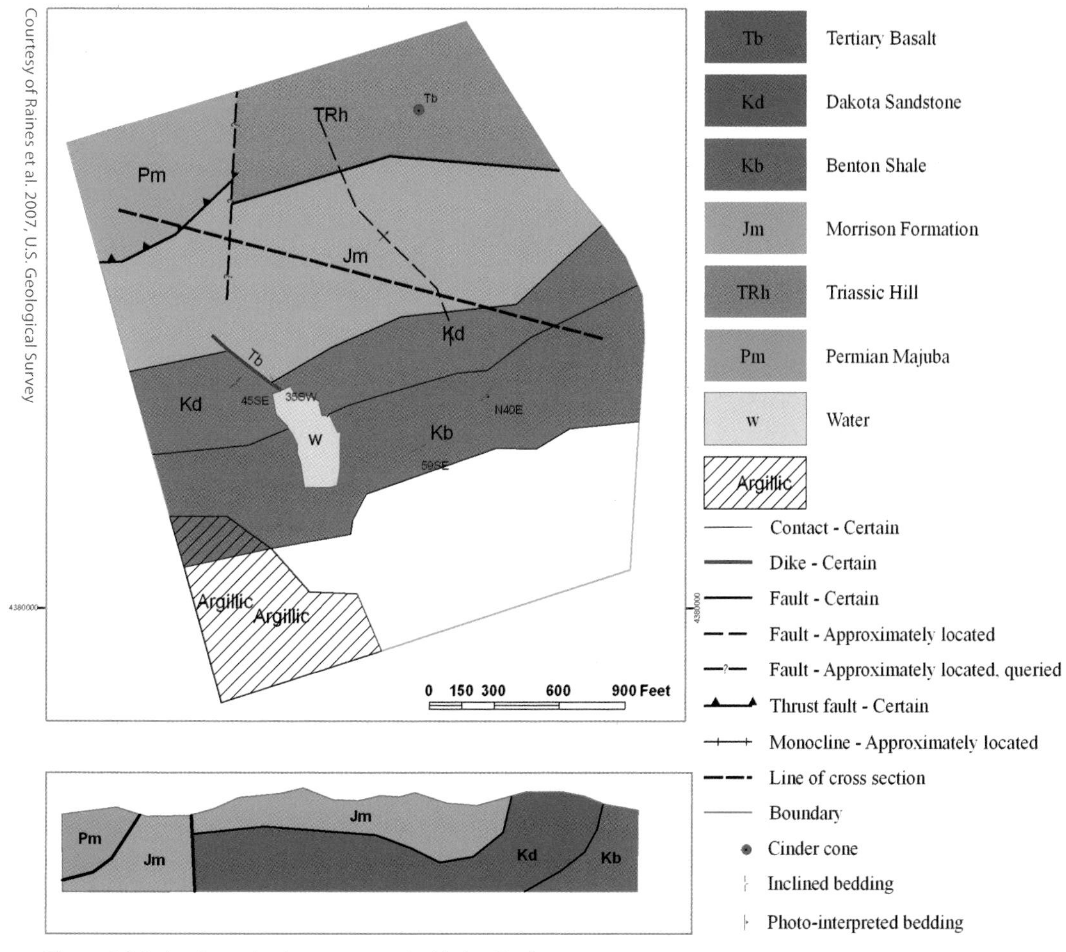

Figure 4.2 A simple geologic map created with Arc Geology.

State agencies have developed project-specific geologic databases to support the production of digital geologic maps and convert existing paper maps into digital format. For example, the USGS-Texas Water Science Center (USGS-TWSC) and Texas Water Development Board have digitized the Geologic Atlas of Texas (GAT), which was developed by the Bureau of Economic Geology at the University of Texas at Austin. The atlas was composed of 38 (1:250,000) hardcopy map sheets, which were scanned and digitized to produce a digital geologic map of Texas (`http://www.tnris.state.tx.us`). To archive the digital maps in a standard form, a GAT geodatabase data model was designed (figure 4.3). Other state geologic surveys have undertaken similar efforts.

Geology component

While it is important to recognize the importance of geologic features for groundwater analysis, we did not attempt to create a comprehensive geologic map database in the groundwater data model design. Instead, we provided a simple placeholder for spatial features that may be included in a groundwater project. The geology component of the data model includes three feature classes: GeologyPoint, GeologyLine, and GeologyArea (figure 4.4). These general classes can be linked to data from more extensive geologic map databases.

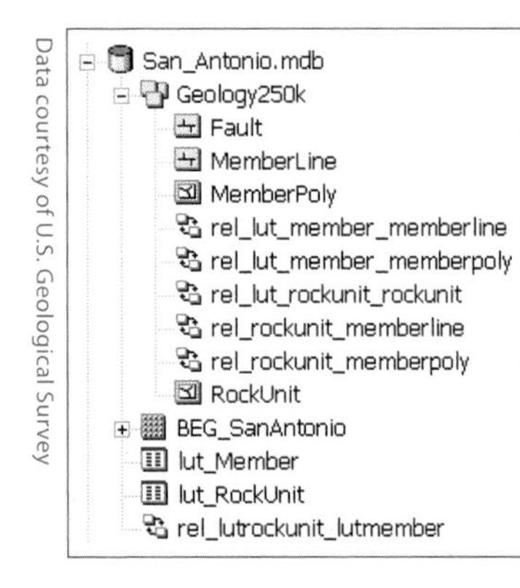

Data courtesy of U.S. Geological Survey

Figure 4.3 Geodatabase design from the Geologic Atlas of Texas for storing geologic map data.

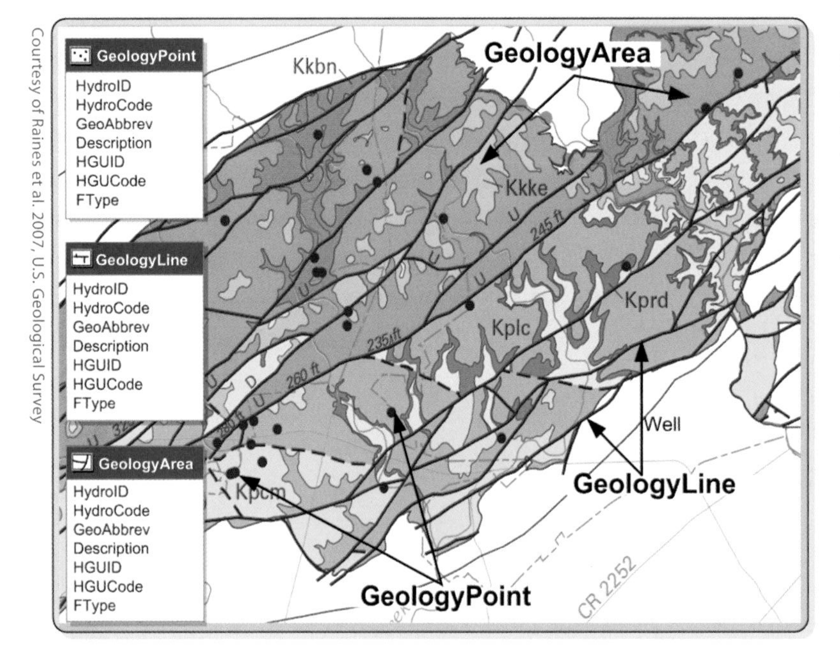

Courtesy of Raines et al. 2007, U.S. Geological Survey

Figure 4.4 Analysis diagram describing the feature classes of the geology component. The background map from (Blome et al. 2005) shows point, line, and polygon features within a geologic map.

GeologyPoints represent point features such as springs, caves, sinks, and observation points. GeologyLines describe line features such as faults, contacts, and dikes. GeologyAreas describe areal features such as rock units and alteration zones.

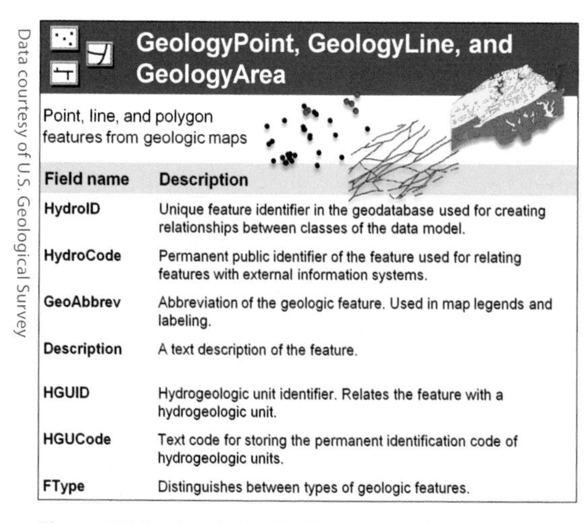

GeologyPoint, GeologyLine, and GeologyArea

Point, line, and polygon features from geologic maps

Field name	Description
HydroID	Unique feature identifier in the geodatabase used for creating relationships between classes of the data model.
HydroCode	Permanent public identifier of the feature used for relating features with external information systems.
GeoAbbrev	Abbreviation of the geologic feature. Used in map legends and labeling.
Description	A text description of the feature.
HGUID	Hydrogeologic unit identifier. Relates the feature with a hydrogeologic unit.
HGUCode	Text code for storing the permanent identification code of hydrogeologic units.
FType	Distinguishes between types of geologic features.

Data courtesy of U.S. Geological Survey

Figure 4.5 GeologyPoint, GeologyLine, and GeologyArea feature classes represent data from geologic maps.

The attributes of the feature classes are described in figure 4.5. The features have HydroID and HydroCode attributes for internal and external identification. GeoAbbrev contains the abbreviation of the unit in the map and map legend, and Description provides a more detailed text description of the unit. HGUID and HGUCode classify the units into hydrogeologic units (see chapter 6 for a more detailed description of the HGUID and HGUCode attributes). FType classifies geologic features by their type (e.g., spring, contact, fault, dike, rock, and alteration).

Geologic maps define the spatial extent, location, and topology of stratigraphic units, which are rocks or bodies of strata recognized as a unit for description, mapping, or correlation purposes. For example, a geologic map of the Edwards Aquifer recharge zone defines fifteen stratigraphic units (figure 4.6). These units can be grouped into more general units (e.g., Kainer Formation and Person Formation), and can also be grouped into aquifers and confining units.

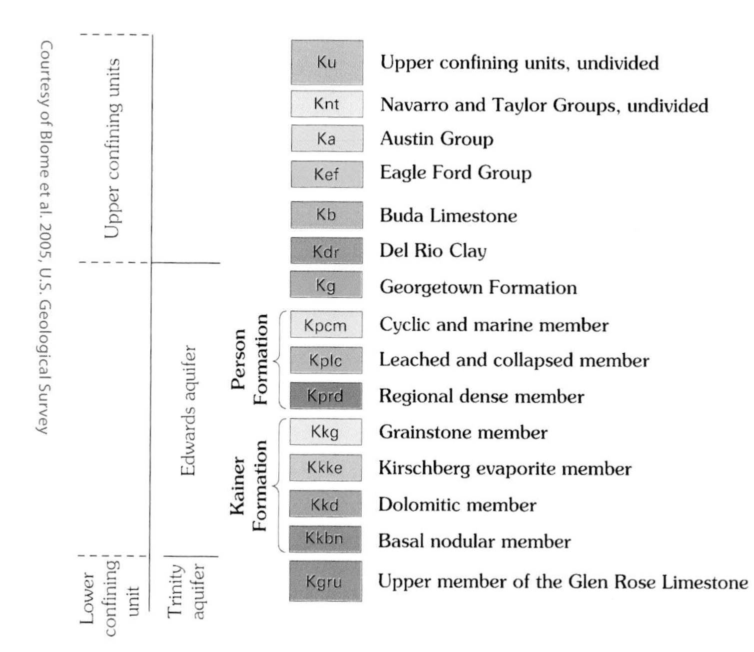

Courtesy of Blome et al. 2005, U.S. Geological Survey

Figure 4.6 Correlation chart showing stratigraphic units, formations, and hydrogeologic units in the Edwards Aquifer recharge zone.

The datasets composing the recharge zone map include polygons describing the extent of geologic members, line features describing faults and contacts, and point features showing the location of caves/sinks and springs (figure 4.7). These features can be used in groundwater projects to help define the extent of hydrogeologic units and provide information on sources (caves/sinks that recharge the aquifer) and discharge features (springs) and flow conduits/barriers (faults).

Implementation

This section outlines the main steps involved in implementing the geology component of the data model. The first steps are to refine the geodatabase design to meet your specific project needs (see chapters 2 and 9 for a detailed description of these steps). Once the geodatabase is built, you can load data into the geology feature classes. Then you can assign HydroIDs to the imported features to

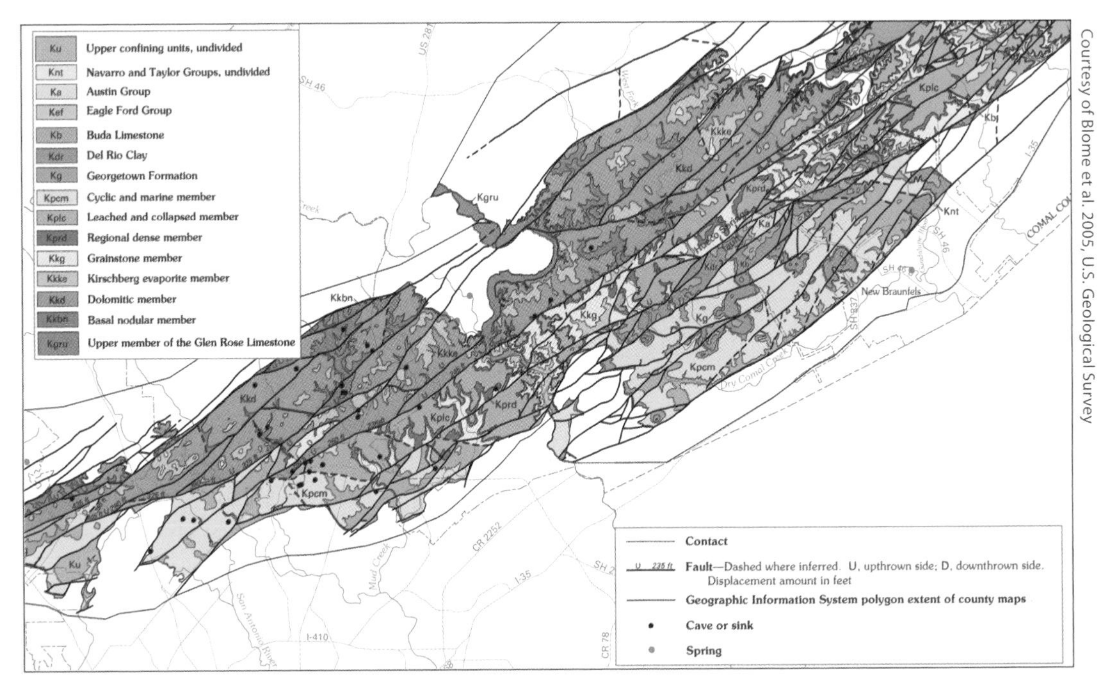

Figure 4.7 Geologic map of the Edwards Aquifer recharge zone. The different colored polygons represent different stratigraphic units, and the points and lines represent faults, springs, and caves or sinks.

create a unique identifier that can be the basis for queries and relationships within the geodatabase. You can then apply tools to the imported data to calculate new attributes and create new features. Finally, you can use the results of this process to create products such as maps, scenes, and reports. The following checklist provides a summary of the main steps in implementing this component.

Checklist

1) Create the classes of the geology component (manually using ArcCatalog or by importing from an XML schema)
2) Define the spatial reference, and add project specific classes, attributes, relationships, and domains as necessary
3) Document datasets and changes made to the data model
4) Import data into the geology classes
5) Assign key attributes to uniquely identify the features and establish relationships
6) Apply tools to create new features and calculate attributes
7) Visualize data and create products (maps, scenes, reports)

The following example demonstrates how data from a USGS geologic map of the Edwards Aquifer recharge zone were incorporated into the geology component of the data model. The GeologyArea feature class was populated with map units from a geologic map database (figure 4.8). After importing the features, each feature was given a HydroID to uniquely identify it within the geodatabase. GeoAbbrev contains the abbreviation of the unit in the map, and Description provides a more detailed text description of the unit. HGUID and HGUCode values were given to the features to classify the members into more general hydrogeologic units. As shown in the map legend in figure 4.9, a number of map members (with different abbreviations) were grouped and classified as more general units (e.g., Person, Kainer, etc.). The FType attribute classifies geologic features by their type. In this case all the units were given a "Map unit" feature type. Where appropriate, FType values can be defined as coded value domains.

Attributes of GeologyArea

SHAPE *	HydroID	HydroCode	GeoAbbrev	Description	HGUID	HGUCode	FType
Polygon	335	<Null>	Kkg	Grainstone member of Kainer Formation (Lower Cretaceous)	3	Kainer	Map Unit
Polygon	336	<Null>	Kkg	Grainstone member of Kainer Formation (Lower Cretaceous)	3	Kainer	Map Unit
Polygon	337	<Null>	Kkg	Grainstone member of Kainer Formation (Lower Cretaceous)	3	Kainer	Map Unit
Polygon	338	<Null>	Kkg	Grainstone member of Kainer Formation (Lower Cretaceous)	3	Kainer	Map Unit
Polygon	339	<Null>	Kkg	Grainstone member of Kainer Formation (Lower Cretaceous)	3	Kainer	Map Unit
Polygon	340	<Null>	Kkg	Grainstone member of Kainer Formation (Lower Cretaceous)	3	Kainer	Map Unit
Polygon	341	<Null>	Kkg	Grainstone member of Kainer Formation (Lower Cretaceous)	3	Kainer	Map Unit
Polygon	342	<Null>	Kkg	Grainstone member of Kainer Formation (Lower Cretaceous)	3	Kainer	Map Unit
Polygon	343	<Null>	Kkbn	Basal nodular member of Kainer Formation (Lower Cretaceous)	3	Kainer	Map Unit
Polygon	344	<Null>	Kkbn	Basal nodular member of Kainer Formation (Lower Cretaceous)	3	Kainer	Map Unit
Polygon	345	<Null>	Kkbn	Basal nodular member of Kainer Formation (Lower Cretaceous)	3	Kainer	Map Unit
Polygon	346	<Null>	Kkbn	Basal nodular member of Kainer Formation (Lower Cretaceous)	3	Kainer	Map Unit
Polygon	347	<Null>	Kkbn	Basal nodular member of Kainer Formation (Lower Cretaceous)	3	Kainer	Map Unit
Polygon	348	<Null>	Kgru	Upper member of Glen Rose Limestone (Lower Cretaceous)	4	Upper Glen Rose	Map Unit
Polygon	349	<Null>	Kgru	Upper member of Glen Rose Limestone (Lower Cretaceous)	4	Upper Glen Rose	Map Unit
Polygon	350	<Null>	Kgru	Upper member of Glen Rose Limestone (Lower Cretaceous)	4	Upper Glen Rose	Map Unit

Record: 0 Show: All Selected Records (0 out of 211 Selected) Options

Figure 4.8 Each row in the GeologyArea feature class represents a map unit.

Similarly, point and line features from the geologic map were imported and attributed in the GeologyPoint and GeologyLine feature classes. In this case, GeologyLine features represent mapped faults, and GeologyPoints represent caves/sinks and springs. As needed, more specific attributes can be added to the feature classes to represent features of interest in more detail. For example, in the map shown in figure 4.9, a FaultType field was added to the GeologyLine features to classify different types of faults (certain or inferred).

References

Geologic Data subcommittee, Federal Geographic Data Committee, FGDC Digital Cartographic Standard for Geologic Map Symbolization. 2006. FGDC Document Number FGDC-STD-013-2006.

Raines, Gary L., Jordan T. Hastings, and Lorre A. Moyer. 2007. Proposed ArcGeology Version 1, A Geodatabase Design for Digital Geologic Maps using ArcGIS. U.S. Geological Survey, Reno, Nevada.

Blome, Charles D., Jason R. Faith, Diana E. Pedraza, George B. Ozuna, James C. Cole, Allan K. Clark, Ted A. Small, and Robert R. Morris. 2005. Geologic map of the Edwards Aquifer recharge zone, south-central Texas. U.S. Geological Survey Special Investigations Map 2873.

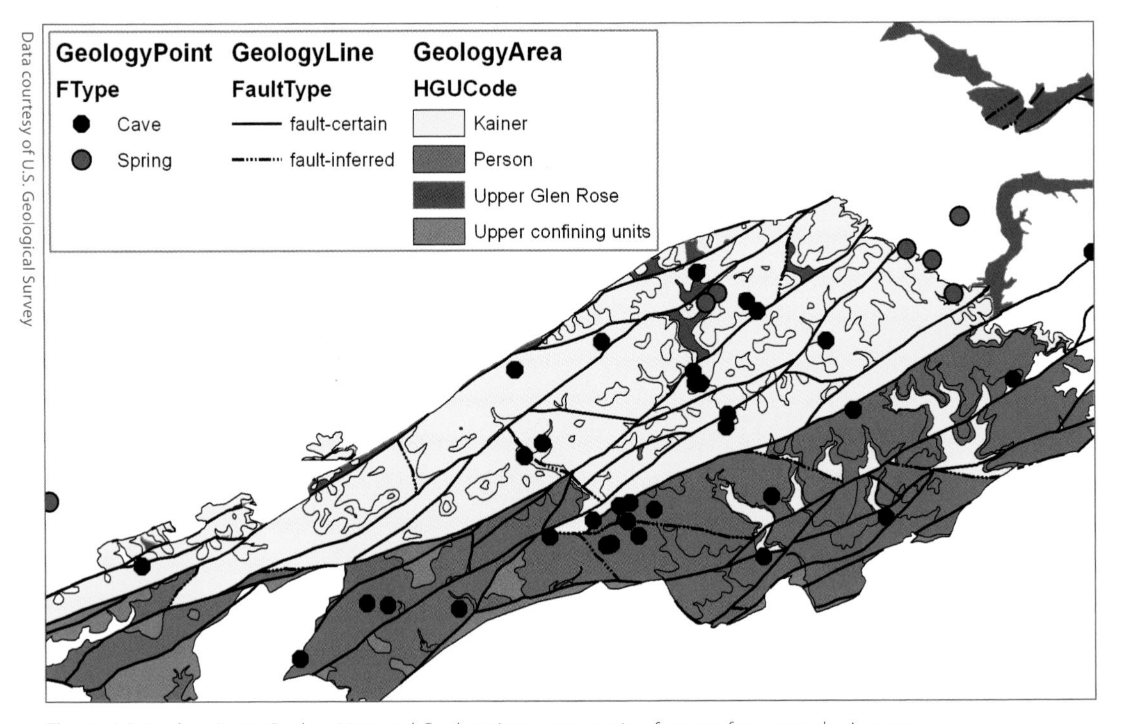

Figure 4.9 GeologyPoint, GeologyLine, and GeologyArea representing features from a geologic map.

Aquifers, wells, and boreholes

GIL STRASSBERG AND NORMAN L. JONES

Data courtesy of U.S. Geological Survey and the Texas Water Development Board

Aquifer
Aquifer boundaries and zones within them such as outcrops and confined zones.
Representation: Polygon features.

Well
Representation of well locations and attributes.
Representation: Point features.

BorePoint and BoreLine
Spatial representation of 3D data along a borehole such as hydrostratigraphy and well construction.
Representation: PointZ and PolylineZ features.

BoreholeLog
Tabular representation of 3D data along a borehole such as hydrostratigraphy and well construction.
Representation: Table.

AQUIFER MAPS AND WELL DATABASES ARE PROBABLY THE MOST COMMON groundwater data products developed on national, state, and local scales, and both serve as important resources for groundwater projects. This chapter shows how the Arc Hydro Groundwater data model represents aquifer and well features and

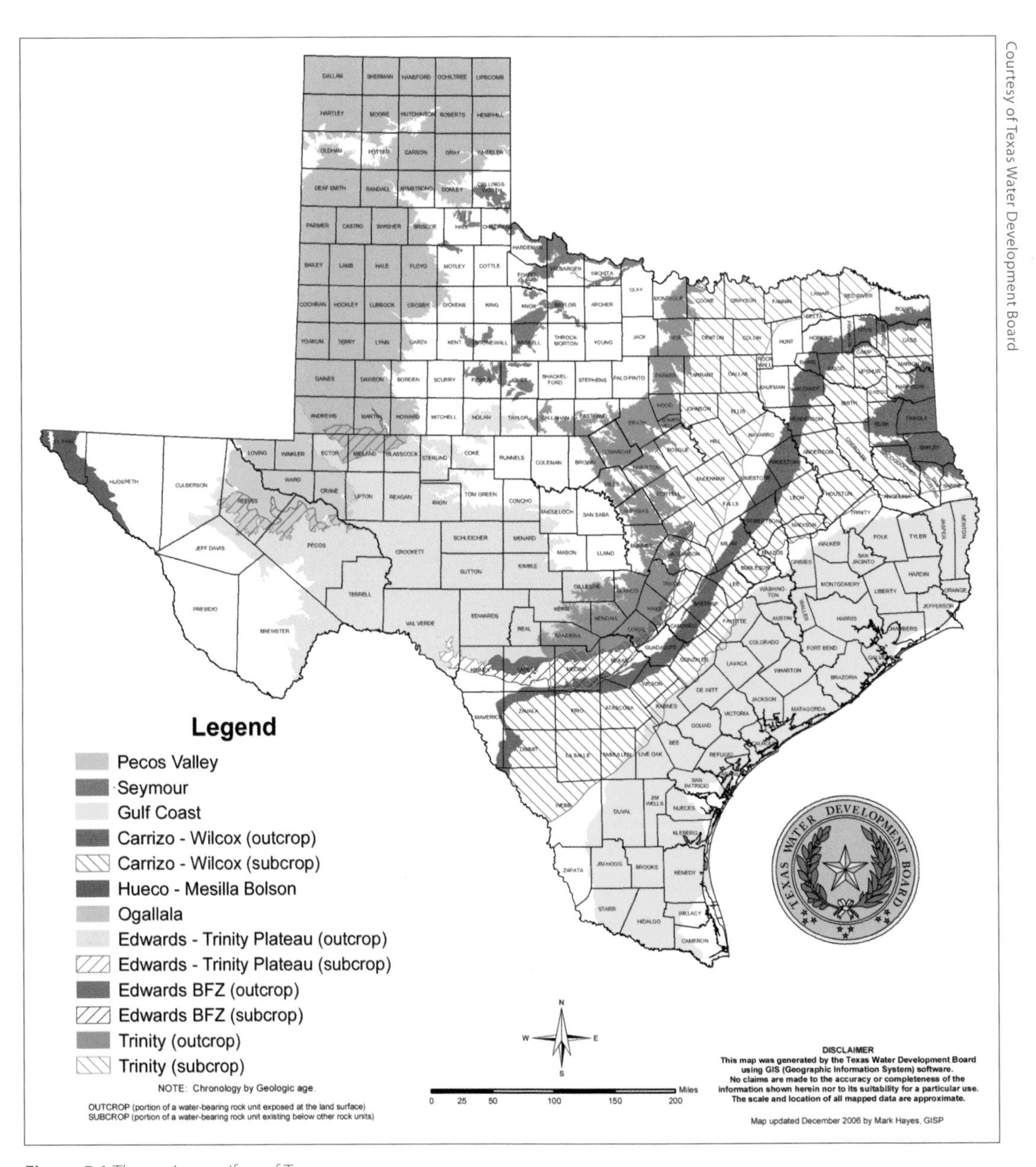

Figure 5.1 The major aquifers of Texas.

how the linkage between the two is established. In addition, the chapter discusses how 3D information recorded along boreholes is stored and visualized.

Aquifer maps and aquifer features

An aquifer is a group of formations, a single geologic formation, or part of a formation that contains sufficient saturated permeable material to yield significant quantities of water to springs and wells (`http://pubs.usgs.gov/gip/gw/glossary.html`). Aquifer maps are common data products developed on national, state, and local levels (figure 5.1). The maps portray the boundary of aquifers over a given region and can also distinguish between zones within an aquifer.

Well databases and well features

A water well is a human-made excavation or structure created in the ground to access groundwater. Wells are among the oldest and most important structures made by humans. Wells have played a crucial role in the development of civilizations since ancient times, as they provide access to groundwater that supports communities even during dry periods when surface water is not available. The importance of wells is documented in numerous literary references, and archeological findings show that wells have supported communities for thousands of years. Wells remain extremely important today, with at least 1.5 billion people worldwide relying on groundwater as their only source of drinking water (`http://www.usaid.gov/our_work/environment/water/groundwater_mgmt.html`). Figure 5.2 shows examples of wells drilled within the Edwards Aquifer.

Wells are designed for many reasons: to produce water from aquifers, to monitor water levels and quality, and to inject (discharge) solutions into the subsurface. Within the groundwater community, the terms well and borehole are commonly used interchangeably. In this book, we refer

Courtesy of Gregg Eckhardt

Figure 5.2 At left, a photograph of two of San Antonio's first municipal artesian wells dug in the Edwards Aquifer shows that the aquifer was under tremendous pressure at that time. At right, a well drilled in the Edwards Aquifer to supply water to a catfish farm (when the well was drilled in the early 1990s, it was considered the largest water well in the world).

to a borehole as a hole drilled into the subsurface, and we refer to a well as a borehole that is related to water extraction, injection, or monitoring.

Cities, counties, and states document and issue permits for well construction. Many of these agencies have established well databases containing information on well locations and well properties describing the construction and usage of wells. These databases provide important information for the development of groundwater-management plans, groundwater simulation models, and research studies dealing with groundwater availability and water quality. Some databases provide detailed information on the well construction, water usage, and hydrostratigraphy, while others contain only basic information such as the owner and location of the wells. Various standards and guidelines have been established for data describing groundwater wells. One is the standard developed by the American Society for Testing and Materials (ASTM), which describes the minimum set of data elements to identify a groundwater site (ASTM 2004). Another is the Australian National Groundwater Data Transfer Standard (Australian National Groundwater Committee 1999), which defines a set of data structures and standards for core groundwater data used in Australia. Commercial software applications (e.g., SiteFX, HydroGeo Analyst, and EQuIS) provide relational database designs to store well-related information and tools to view and analyze the data. The Arc Hydro Groundwater data model is not intended to replace these standards and applications but to provide geospatial classes for storing, viewing, and analyzing these data within ArcGIS. Rather than trying to be all inclusive and develop a new comprehensive standard, we asked ourselves, what are the common attributes in these well databases, and how can we best store, view, and analyze them within a GIS? This approach provides a simple template that users can customize as needed for their particular project.

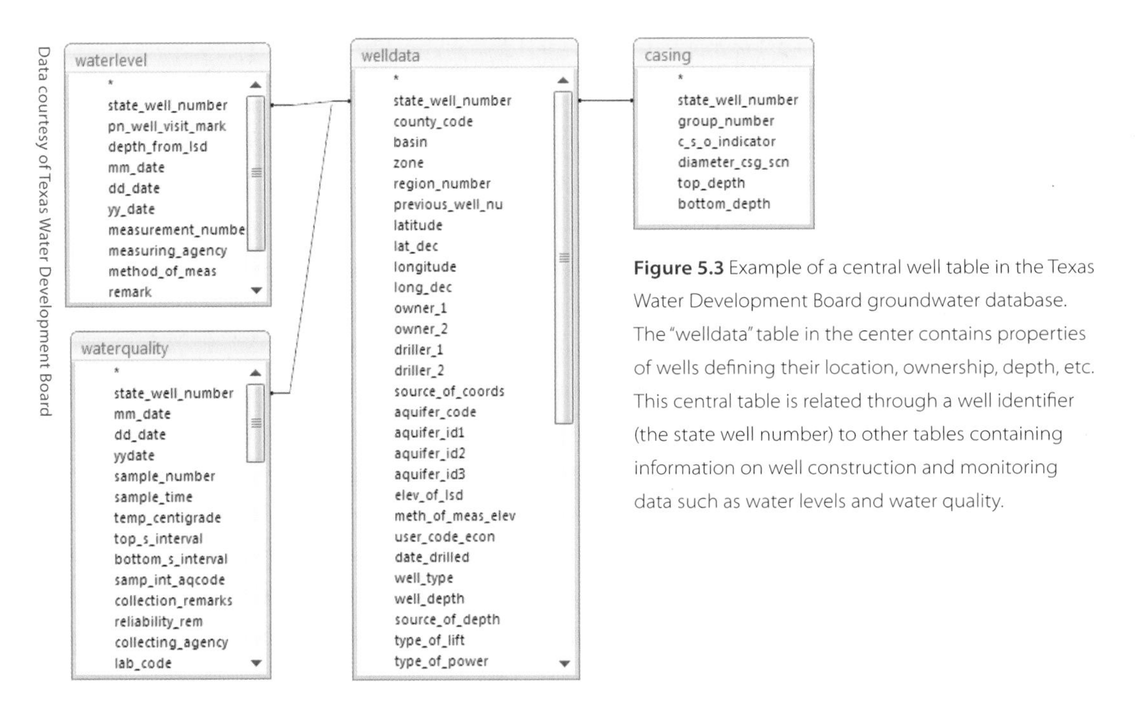

Figure 5.3 Example of a central well table in the Texas Water Development Board groundwater database. The "welldata" table in the center contains properties of wells defining their location, ownership, depth, etc. This central table is related through a well identifier (the state well number) to other tables containing information on well construction and monitoring data such as water levels and water quality.

It is common for groundwater databases to have a central table that describes the location of wells and their properties (e.g., depth, water use, and owner). The central table is usually linked with additional tables describing information on the well construction, measurements of water levels and water quality taken at the well, and descriptions of the formation material observed along boreholes. For example, the groundwater database maintained by the Texas Water Development Board (TWDB) is available for public use and downloadable as a Microsoft Access database from the TWDB Web site. The main table in the database is the "welldata" table (figure 5.3). Each row in the table represents a single well, and fields in the table describe the well's properties, including its geographic coordinates, land surface elevation, depth, owner, driller, and related aquifers. Each well in the TWDB groundwater database is assigned a "state well number" that identifies the well within the database. Based on this number, associations are created to relate wells in the welldata table with data in other tables containing information on the casing intervals along the well, and on the water quality and water level sample data.

Data from the well database can be visualized in a map environment by representing wells as point features, and the points can be the basis for querying and displaying well-related data. A good example of such a system is the Texas Water Information Integration & Dissemination (WIID) system. The WIID system provides a Web-based mapping interface for retrieving well data from drilling reports and groundwater monitoring data. When selecting a well in the map, basic information on the well is provided. This includes the well's owner, water use, location, elevation, depth, and associated aquifer (figure 5.4). Upon request, users can retrieve data on water quality, water levels, well casing, and the driller report for the well. Similar systems have been developed for

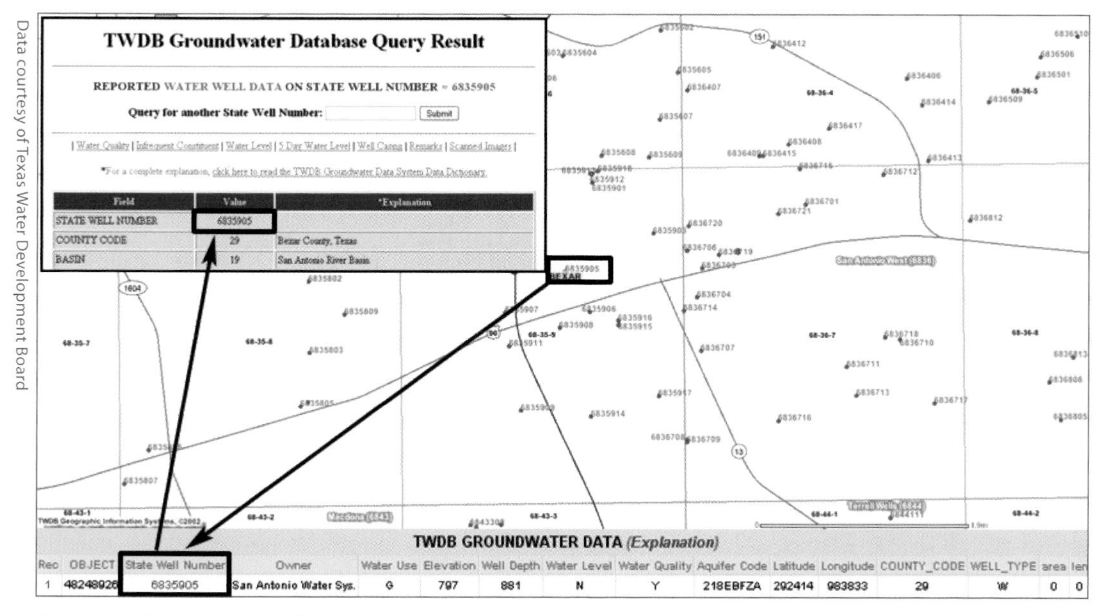

Figure 5.4. The Texas Water Information Integration & Dissemination system (http://wiid.twdb.state.tx.us) is an example of a Web-based mapping interface for retrieving well data from a groundwater database.

many U.S. states (e.g., Michigan, Washington, Illinois) and on the national level (e.g., the USGS National Water Information System).

A theme that emerges from these information systems is the development of a 2D spatial representation of wells (points on a map) to which other data are related. Thus, the first step in spatially enabling well databases is to create a spatial representation of well features. In the groundwater data model, wells are represented as 2D points with attributes describing the properties of the well. Related data, such as well construction, hydrostratigraphy, and monitoring information, are associated with well features. The data model provides a basic set of well attributes for identification, 3D representation, and linkages to aquifer and hydrogeologic units. While many groundwater databases contain large sets of fields describing wells, the groundwater data model focuses on providing a minimum set of attributes that help integrate the well data with GIS. This set of attributes can be expanded for each project as needed.

Geodatabase representation of aquifers and wells

The Aquifer feature class is a polygon feature class for representing data from aquifer maps. Each Aquifer feature in the feature class represents an aquifer boundary or part of an aquifer (see figure 2.10 for a detailed description of the attributes of the Aquifer feature class). The HydroID and HydroCode attributes provide the means for internal and external identification (figure 5.5). For example, the major aquifers of Texas have a unique identifier used by Texas state agencies. The aquifer identifier of the Edwards Aquifer is 11, thus when creating aquifer features, the HydroCode of

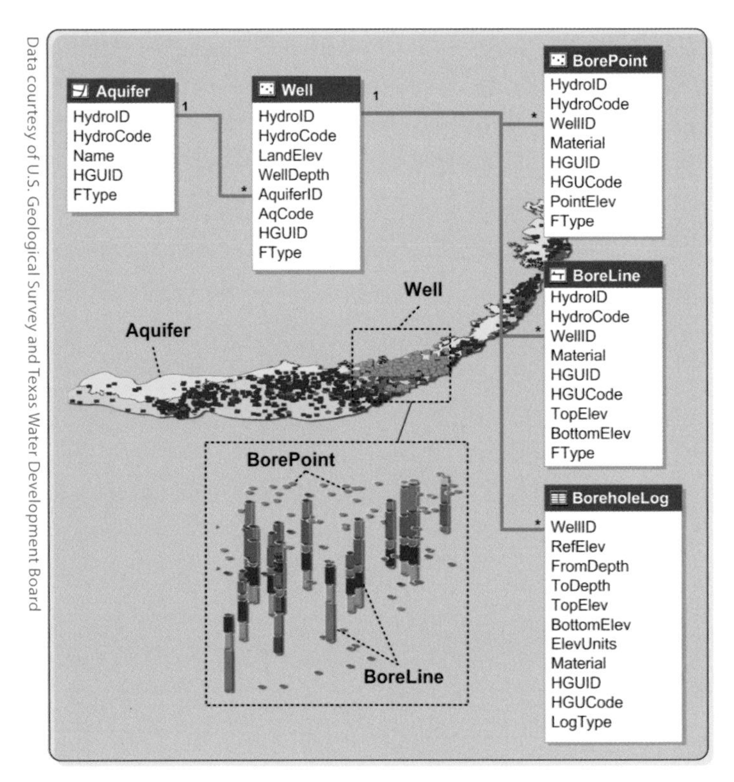

Figure 5.5. Analysis diagram describing the datasets for representing aquifers, wells, and boreholes.

features representing the Edwards Aquifer is set to 11. This lets users link Aquifer features in an Arc Hydro geodatabase with data from external information systems. The Name attribute is used to symbolize and label features, and the HGUID (hydrogeologic unit ID) relates aquifer features with more detailed descriptions of hydrogeologic units defined in the HydrogeologicUnit table (see chapter 6 for a more detailed description of hydrogeologic units). HGUID is also used to group aquifer features. For example, figure 5.6 shows a dataset describing the Edwards Aquifer that is composed of a number of polygon features (each with a unique HydroID). The features in the example are indexed by HGUID = 4 (which is defined in the HydrogeologicUnit table as the boundary of the Edwards Aquifer). Thus, the HGUID can be used to group, display, or query the features representing the Edwards Aquifer. In the following example features are also symbolized by the FType to differentiate between outcrop and confined zones.

The Well feature class is a point feature class for representing well locations and basic well attributes for identification, 3D representation, and linkages to aquifer and hydrogeologic units (see figure 2.11 for a detailed description of the attributes of the Well feature class). HydroID and HydroCode are used for internal and external identification of the features. Each Well feature is assigned a HydroID that becomes its unique identifier within the geodatabase. The HydroID is the

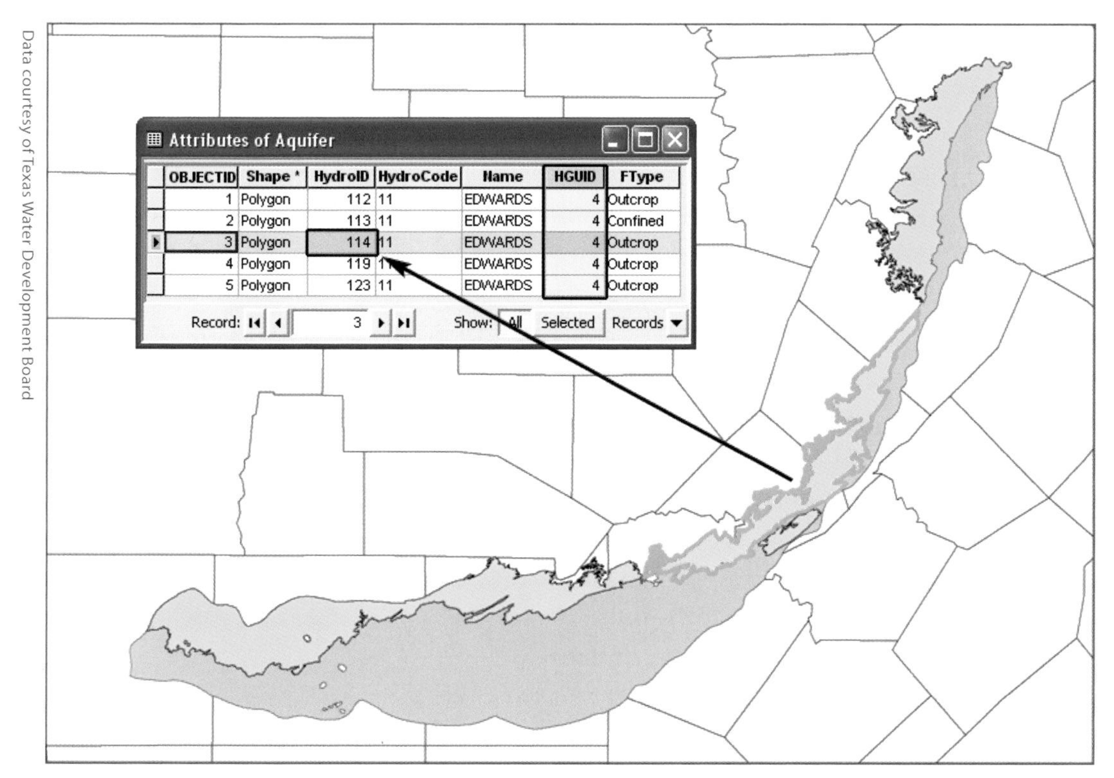

Figure 5.6. Polygon features representing the Edwards Aquifer are indexed by HGUID so that a group of features represents a single aquifer.

basis for relating data in other feature classes and tables to Well features. The HydroCode is the permanent public identifier that helps maintain the original identification code of the well. For example, HydroCode can contain the state well number from a state groundwater database. The LandElev (land elevation) and WellDepth (well depth) attributes describe the 3D geometry of the well, and AquiferID (Aquifer identifier) and AqCode (Aquifer Code) associate wells with aquifers. HGUID relates the well to hydrogeologic units defined in the HydrogeologicUnit table. Figure 5.7 shows an example of Well features in the Edwards Aquifer created by importing data from the TWDB groundwater database.

Relationships between Aquifer and Well features

In the groundwater data model, Well and Aquifer features are related to provide the capabilities to query and display information regarding specific aquifers. The association between Aquifer and Well features is a one-to-many (1:M) relationship, thus an Aquifer feature can be associated with one or more Well features. The relationship is based on key fields in the Aquifer and Well feature classes, such that the AquiferID of Well features is set equal to the HydroID of an Aquifer feature. In the following example (figure 5.8), a Well feature representing a public water supply well is attributed with an

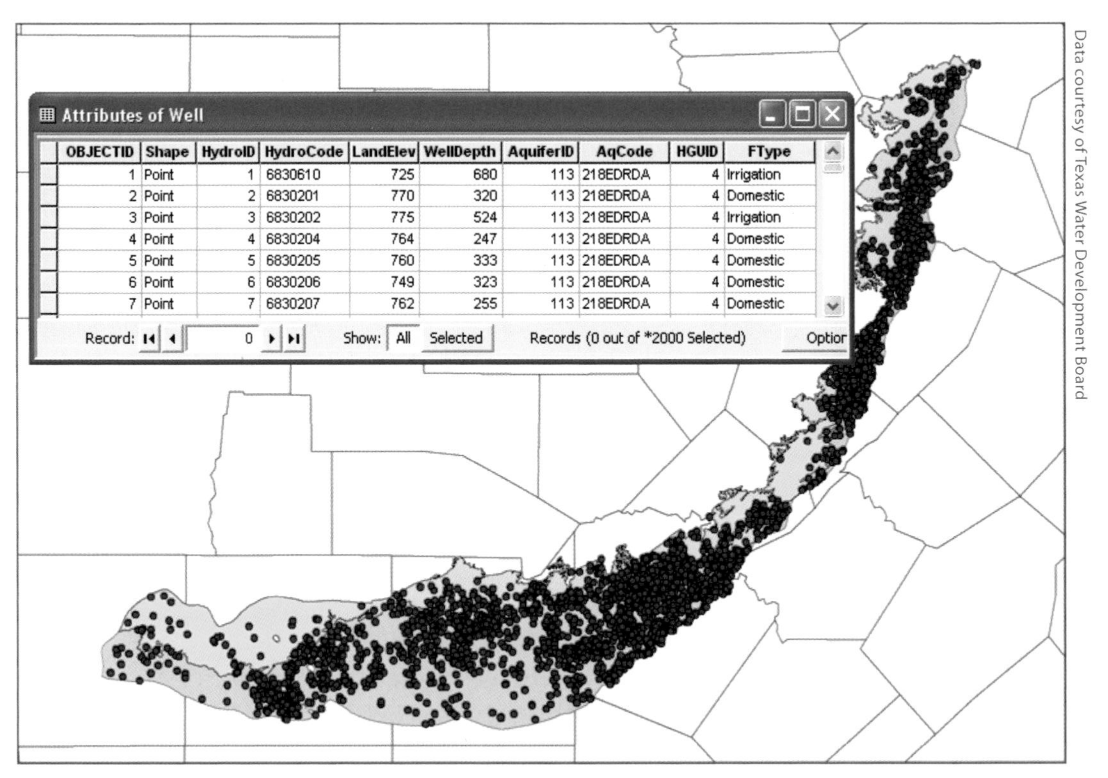

Attributes of Well

OBJECTID	Shape	HydroID	HydroCode	LandElev	WellDepth	AquiferID	AqCode	HGUID	FType
1	Point	1	6830610	725	680	113	218EDRDA	4	Irrigation
2	Point	2	6830201	770	320	113	218EDRDA	4	Domestic
3	Point	3	6830202	775	524	113	218EDRDA	4	Irrigation
4	Point	4	6830204	764	247	113	218EDRDA	4	Domestic
5	Point	5	6830205	760	333	113	218EDRDA	4	Domestic
6	Point	6	6830206	749	323	113	218EDRDA	4	Domestic
7	Point	7	6830207	762	255	113	218EDRDA	4	Domestic

Record: 0 Show: All Selected Records (0 out of *2000 Selected) Optior

Figure 5.7 Well features representing well locations and attributes of wells in the Edwards Aquifer.

AquiferID of 114. This relates the well to an Aquifer feature with HydroID 114, which is one of the features representing the Edwards Aquifer outcrop. The association between Aquifer and Well features enables querying well-related data for specific aquifers. Well features also can be labeled and symbolized by their related aquifer using the aquifer identifier (AquiferID) and code (AqCode) attributes.

In some cases wells are associated with more than one aquifer. The assignment of multiple aquifers can result from a well being screened across multiple aquifer units or from the assignment of aquifers mapped at different scales. For example, wells stored in the USGS National Water Information System (NWIS) are indexed with a national aquifer and a local aquifer. To support the association with both, we can simply add another aquifer identifier attribute to the Well feature class (e.g., AquiferID2) and add another relationship that associates the new aquifer identifier with the HydroID of Aquifer features. This pattern can be repeated to create as many aquifer-well relationships as needed in your project. You can also create a many-to-many (M:N) association between the Aquifer and Well feature classes (our experience shows that managing an M:N relationship is more challenging from a user perspective). If your datasets include a many-to-many relationship between aquifer and wells and it is important for your project to support this relationship, you should consider using one of the approaches described above to represent such an association (see also chapter 9).

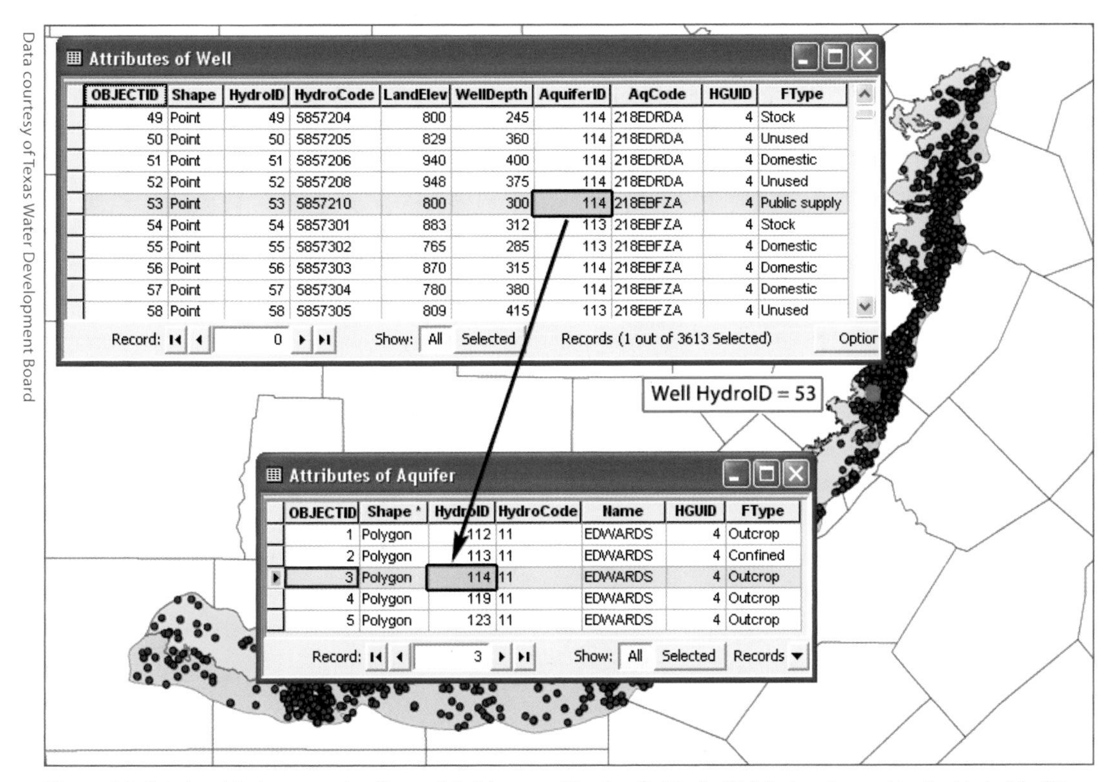

OBJECTID	Shape	HydroID	HydroCode	LandElev	WellDepth	AquiferID	AqCode	HGUID	FType
49	Point	49	5857204	800	245	114	218EDRDA	4	Stock
50	Point	50	5857205	829	360	114	218EDRDA	4	Unused
51	Point	51	5857206	940	400	114	218EDRDA	4	Domestic
52	Point	52	5857208	948	375	114	218EDRDA	4	Unused
53	Point	53	5857210	800	300	114	218EBFZA	4	Public supply
54	Point	54	5857301	883	312	113	218EBFZA	4	Stock
55	Point	55	5857302	765	285	113	218EBFZA	4	Domestic
56	Point	56	5857303	870	315	114	218EBFZA	4	Domestic
57	Point	57	5857304	780	380	114	218EBFZA	4	Domestic
58	Point	58	5857305	809	415	113	218EBFZA	4	Unused

OBJECTID	Shape *	HydroID	HydroCode	Name	HGUID	FType
1	Polygon	112	11	EDWARDS	4	Outcrop
2	Polygon	113	11	EDWARDS	4	Confined
3	Polygon	114	11	EDWARDS	4	Outcrop
4	Polygon	119	11	EDWARDS	4	Outcrop
5	Polygon	123	11	EDWARDS	4	Outcrop

Figure 5.8. Relationship between Aquifer and Well features. The AquiferID of a Well feature is equal to the HydroID of the related Aquifer feature.

Steps for creating Aquifer and Well features

The first steps for implementing this component are to refine the geodatabase design to meet your specific project needs (see chapters 2 and 9 for a detailed description of these steps). For example, you may want to create a list of coded value domains for the FType attribute to constrain well types entered to the feature class or modify the relationship between Aquifer and Well features to fit your project datasets. After building the geodatabase, you can load data into the feature classes, and then assign HydroIDs to the imported features to create a unique identifier that will be the basis for queries and relationships within the geodatabase. You can then populate the key fields in the feature classes (e.g., AquiferID) to establish relationships and apply tools to the imported data to calculate new attributes and create new features. Finally, you can use the results of this process to create products such as maps, scenes, and reports. The following checklist provides a summary of the main steps for creating Aquifer and Well features.

Checklist

1. Create the Aquifer and Well feature classes (manually or by importing from an XML schema)
2. Add project specific attributes and domains as necessary (e.g., aquifer type, well type)
3. Document datasets and changes made to the data model
4. Define the data to be imported (all data or data for specific aquifers, counties, etc.)
5. Load data into the feature classes (if wells are defined in tabular format you first need to create point features from x,y coordinates)
6. Assign unique identifiers to features
7. Populate the relationship between Aquifer and Well features
8. Apply tools (e.g., calculate additional attributes, digitize new features)
9. Create products (maps, scenes, reports)

The following example illustrates the process of creating Aquifer and Well features. Aquifer features were imported into the Arc Hydro Groundwater geodatabase from a dataset created by the TWDB describing all the major aquifers in Texas. Because we are primarily interested in the Edwards Aquifer, only features with an Aquifer ID = 11 (the aquifer identifier in the TWDB database representing the Edwards Aquifer) were imported into the geodatabase. HydroIDs were assigned to the features to uniquely identify them within the geodatabase, and HydroCode was set to 11 to store the original aquifer identifier. Aquifer features were also given an FType to distinguish between confined and outcrop zones (see figure 5.6).

Well features were created by importing information from the "welldata" table of the TWDB groundwater database. The table contains a description of about 130,000 wells in Texas, and each well is described by a set of about 50 attributes. The first step was to define a set of attributes of interest that will be valuable for the project. One could just import all 50 attributes from the TWDB table, but many of these may not be populated or necessary. As described above, the groundwater data model provides a minimum set of attributes, and users must decide what additional information is useful for their project and add these fields to the Well feature class. In the example shown in figure 5.9, fields describing the owner and the water use were added. Later, a set of coded value domains are defined for the water use field representing types of wells within the project.

Once the feature classes are designed, we can load data into them, but first we need to decide on the data to be imported. Will we import data for all the wells in Texas, or are we interested only in the data for a particular county, groundwater management district, or aquifer? In this example, we use only wells in the Edwards Aquifer. In order to import features into the Well feature class, we first create a spatial representation of the wells defined by the coordinates in the well table (this can be done in ArcGIS with the Make XY Event Layer geoprocessing tool). Then the features are loaded into the Well feature class. When loading the data, we only import attributes of interest and only import wells in our area of interest, the Edwards Aquifer (AquiferID = 11).

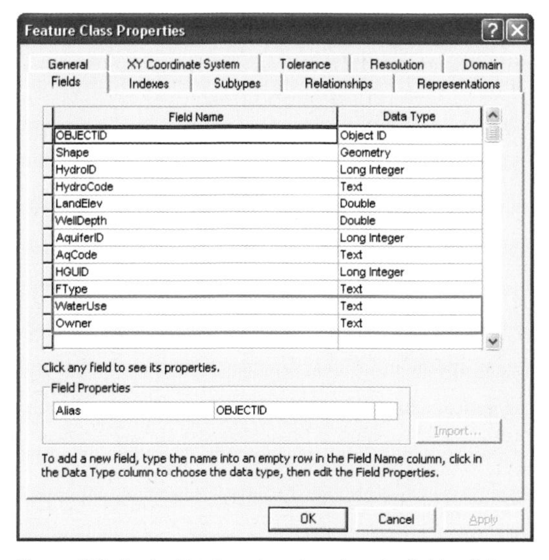

Figure 5.9. An ArcCatalog view showing the fields of the Well feature class. The highlighted fields are not part of the core data model and are added as needed.

After importing the well data and creating Well features, we assign HydroIDs (using the Assign HydroID tool in the Arc Hydro tools). During this import process, the state well number is imported into the HydroCode attribute to establish a connection with the original database from which the data were created. Also, the HGUID is populated to index Well features with a specific hydrogeologic unit, and the aquifer-well relationship is established by indexing Well features with the HydroID of an Aquifer feature.

The next step is to define types of wells represented in the geodatabase. FType can be defined using a coded value domain. For example a WellFType coded value domain is created to define a set of unique values that can be entered in the FType field. The coded value domains can be edited, deleted, and extended to fit your project requirements (See chapter 9 for more information on creating coded value domains). After assigning the coded value domain, when editing the FType field in ArcMap, values will be restricted to the set of coded values defined in the Well FType domain. In an edit session, a drop-down menu appears when you try to edit the FType values, and you must select one of the predefined coded values (figure 5.10).

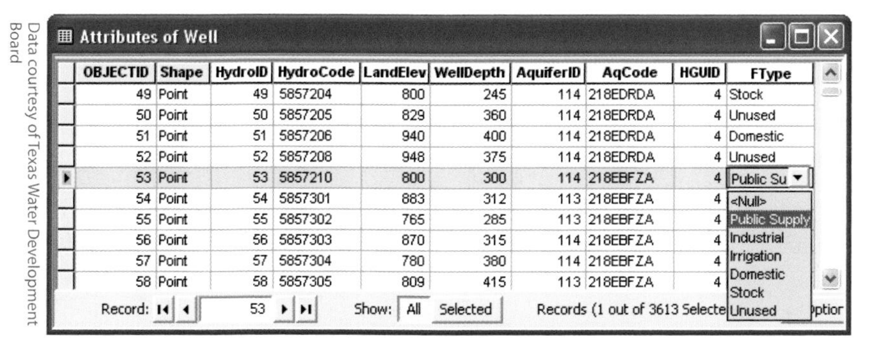
Attributes of Well

OBJECTID	Shape	HydroID	HydroCode	LandElev	WellDepth	AquiferID	AqCode	HGUID	FType
49	Point	49	5857204	800	245	114	218EDRDA	4	Stock
50	Point	50	5857205	829	360	114	218EDRDA	4	Unused
51	Point	51	5857206	940	400	114	218EDRDA	4	Domestic
52	Point	52	5857208	948	375	114	218EDRDA	4	Unused
53	Point	53	5857210	800	300	114	218EBFZA	4	Public Su
54	Point	54	5857301	883	312	113	218EBFZA	4	
55	Point	55	5857302	765	285	113	218EBFZA	4	
56	Point	56	5857303	870	315	114	218EBFZA	4	
57	Point	57	5857304	780	380	114	218EBFZA	4	
58	Point	58	5857305	809	415	113	218EBFZA	4	

Data courtesy of Texas Water Development Board

Figure 5.10. When editing an attribute with a coded value domain the values entered are restricted to a predefined list.

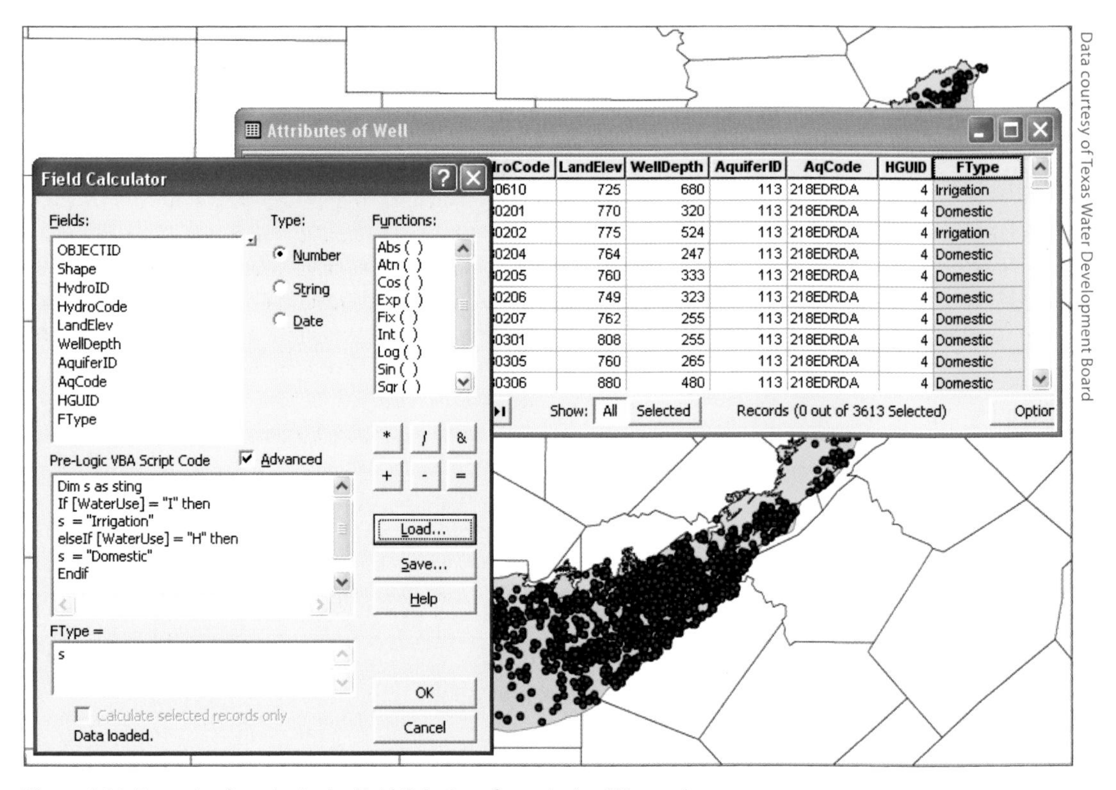

Figure 5.11. Example of a script in the Field Calculator for assigning FType values.

In the Texas groundwater database, water use was defined using single letters. For example, "Water Use = I" stands for irrigation, and "Water Use = H" stands for domestic. Figure 5.11 shows a simple script written in the Field Calculator that is useful for automating the assignment of FType values.

At the end of this process, tabular data from the TWDB groundwater database were converted to spatial features representing wells in the Edwards Aquifer, and these were imported into the Well feature class. The features are attributed such that they can be linked with related Aquifer features, and are also classified by the Well FType coded value domain. The following map (figure 5.12) shows Well features in the Edwards Aquifer classified by the FType.

3D borehole data

Although wells in the data model are represented as 2D point features, in reality the geometry of a well is 3D as it extends into the subsurface. The 3D geometry of a Well feature is defined by the x and y coordinates of the point feature and in the vertical (z) dimension by the LandElev and WellDepth attributes. The 3D geometry of wells can be viewed in ArcScene by extruding the Well point from the land surface downward (figure 5.13).

Many groundwater and well databases include tables for describing 3D construction elements (e.g., screens, casing, and samplers) and tables for describing strata along boreholes. These types of data are commonly documented in construction and drilling logs and are later stored as tabular information in groundwater and well databases. In many cases these datasets follow a common data structure, where each record corresponds to a set of elevations representing vertical segments along the borehole. The following section describes how such data are represented within the groundwater data model. Throughout this section we use the term borehole for describing the vertical dimension of a well.

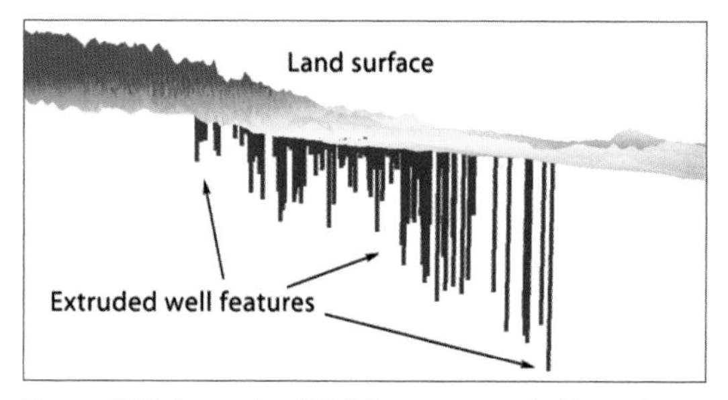

Figure 5.13. Example of Well features extruded from the land surface to the well depth.

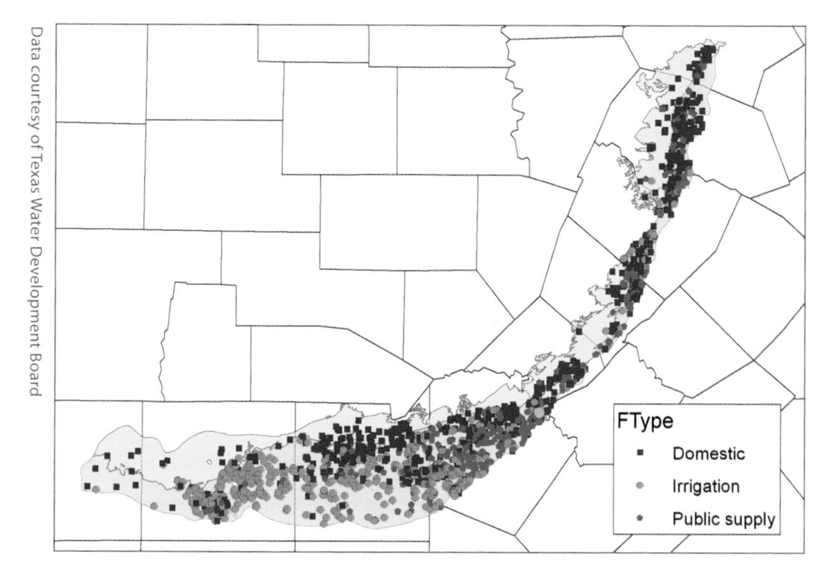

Figure 5.12. Well features representing wells in the Edwards Aquifer. The features are classified by FType to distinguish among domestic, irrigation, and public supply wells.

BoreholeLog table

BoreholeLog is a table for representing vertical data along boreholes and is the basis for creating 3D features to represent vertical data as 3D geometries. Data in the BoreholeLog table do not contain x and y coordinates; rather a Well feature is referenced by indexing data in the table with a WellID attribute that is equal to the HydroID of a Well feature. This structure implies that the borehole represented is vertical and that all objects along the borehole have the same x and y coordinates as the referenced Well feature. The association between Well features and BoreholeLog objects is a one-to-many (1:M) relationship, where a Well feature can be associated with one or more vertical data elements in the BoreholeLog table. Figure 5.14 shows the structure of the BoreholeLog table.

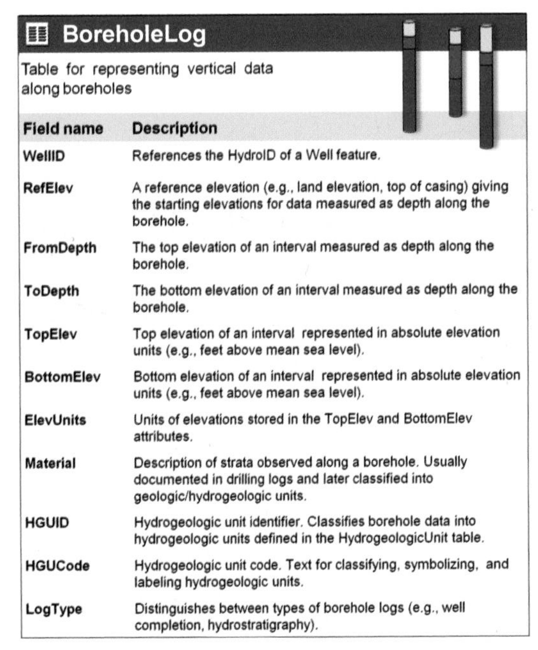

BoreholeLog

Table for representing vertical data along boreholes

Field name	Description
WellID	References the HydroID of a Well feature.
RefElev	A reference elevation (e.g., land elevation, top of casing) giving the starting elevations for data measured as depth along the borehole.
FromDepth	The top elevation of an interval measured as depth along the borehole.
ToDepth	The bottom elevation of an interval measured as depth along the borehole.
TopElev	Top elevation of an interval represented in absolute elevation units (e.g., feet above mean sea level).
BottomElev	Bottom elevation of an interval represented in absolute elevation units (e.g., feet above mean sea level).
ElevUnits	Units of elevations stored in the TopElev and BottomElev attributes.
Material	Description of strata observed along a borehole. Usually documented in drilling logs and later classified into geologic/hydrogeologic units.
HGUID	Hydrogeologic unit identifier. Classifies borehole data into hydrogeologic units defined in the HydrogeologicUnit table.
HGUCode	Hydrogeologic unit code. Text for classifying, symbolizing, and labeling hydrogeologic units.
LogType	Distinguishes between types of borehole logs (e.g., well completion, hydrostratigraphy).

Figure 5.14. The BoreholeLog table represents vertical information recorded along boreholes.

Features along the borehole can be described as points defined by a single elevation or as intervals defined by two elevations, one for the top and one for the bottom of the interval. Top and bottom elevations can be defined with two types of referencing systems: depth along the borehole (relative elevations) or absolute elevations. Relative elevations are measured from a reference elevation, which is usually the land surface at the well location (figure 5.15). This is a common method for storing information from borehole logs. Usually, data are recorded during the drilling and construction process and are defined as "depth below the land surface." Thus, the vertical coordinate increases in the downward direction. The depth values are stored in the FromDepth and ToDepth attributes. When using such a referencing system, vertical data from different wells cannot be directly compared because the elevation of the land surface is not the same at each well location. To integrate data across a number of wells (e.g., interpolate rasters, create cross sections) absolute elevations must be used. Absolute elevations are referenced to a common datum, such as mean sea

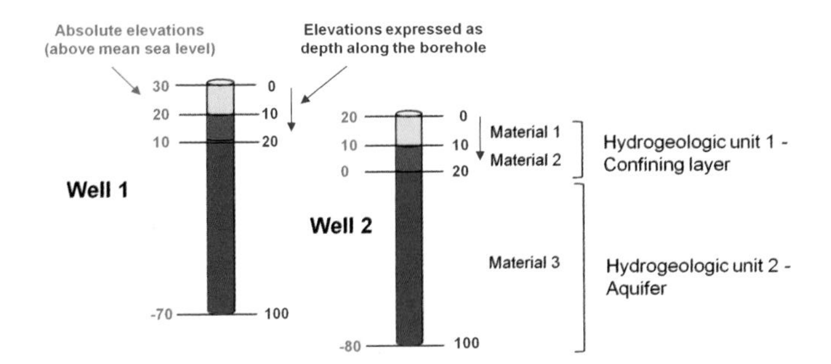

Figure 5.15 Vertical data along boreholes are represented using absolute elevations (in green) and as depth along the borehole (in blue). Data are also classified by material, and materials can be grouped into hydrogeologic units.

level, and can be directly compared. Usually, raw data are given as depth along the borehole, and these are translated to absolute elevations. The absolute values are stored in the TopElev and BottomElev attributes.

Figure 5.16 shows a populated BoreholeLog table for the boreholes described in figure 5.15. The FromDepth and ToDepth values for both wells are the same, but because the reference elevation (RefElev) is different, the absolute elevations (stored as feet above mean sea level) represented in the TopElev and BotmElev attributes vary accordingly.

BorePoint and BoreLine features

Data stored in the BoreholeLog table are the basis for creating 3D features that represent vertical data along boreholes. BorePoint and BoreLine are 3D (z-enabled) point and line feature classes that represent point and interval data along boreholes, respectively. Similar to the relationship between BoreholeLog objects and Well features, BorePoint and BoreLine features are related to a Well feature by associating the WellID attribute of the 3D features to the HydroID of a Well feature. TopElev, BottomElev, and PointElev are attributes for storing the elevations of the features. While the elevations are already stored internally in the geometry of the features, storing them as attributes enables easy display of elevations and can support elevation-based queries. For example, we could query for all BorePoints with an elevation between 500 and 800 feet above mean sea level. Figure 5.17 shows the structure of the BorePoint and BoreLine feature classes.

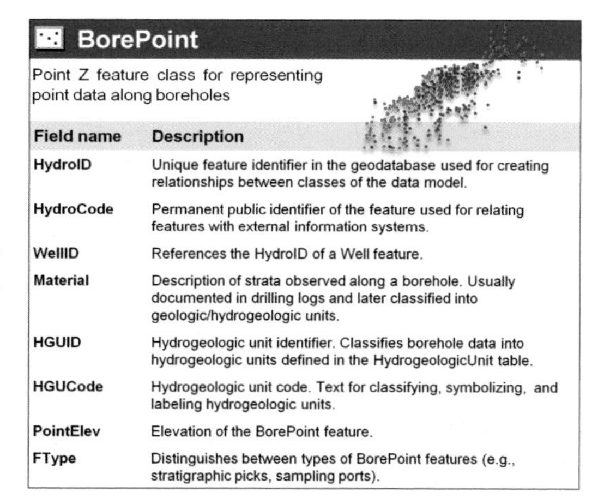

BorePoint

Point Z feature class for representing point data along boreholes

Field name	Description
HydroID	Unique feature identifier in the geodatabase used for creating relationships between classes of the data model.
HydroCode	Permanent public identifier of the feature used for relating features with external information systems.
WellID	References the HydroID of a Well feature.
Material	Description of strata observed along a borehole. Usually documented in drilling logs and later classified into geologic/hydrogeologic units.
HGUID	Hydrogeologic unit identifier. Classifies borehole data into hydrogeologic units defined in the HydrogeologicUnit table.
HGUCode	Hydrogeologic unit code. Text for classifying, symbolizing, and labeling hydrogeologic units.
PointElev	Elevation of the BorePoint feature.
FType	Distinguishes between types of BorePoint features (e.g., stratigraphic picks, sampling ports).

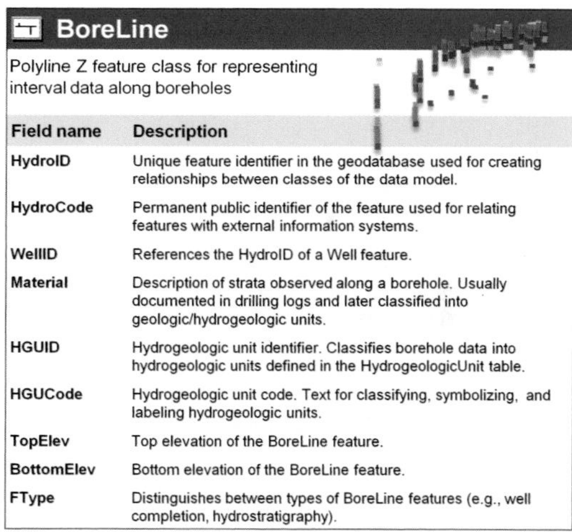

BoreLine

Polyline Z feature class for representing interval data along boreholes

Field name	Description
HydroID	Unique feature identifier in the geodatabase used for creating relationships between classes of the data model.
HydroCode	Permanent public identifier of the feature used for relating features with external information systems.
WellID	References the HydroID of a Well feature.
Material	Description of strata observed along a borehole. Usually documented in drilling logs and later classified into geologic/hydrogeologic units.
HGUID	Hydrogeologic unit identifier. Classifies borehole data into hydrogeologic units defined in the HydrogeologicUnit table.
HGUCode	Hydrogeologic unit code. Text for classifying, symbolizing, and labeling hydrogeologic units.
TopElev	Top elevation of the BoreLine feature.
BottomElev	Bottom elevation of the BoreLine feature.
FType	Distinguishes between types of BoreLine features (e.g., well completion, hydrostratigraphy).

Figure 5.17 BorePoint and BoreLine feature classes represent vertical data along boreholes as 3D features.

Attributes of BoreholeLog

OBJECTID	WellID	RefElev	FromDepth	ToDepth	TopElev	BottomElev	ElevUnits	Material	HGUID	HGUCode	LogType
1	1	30	0	10	30	20	Feet above mean sea level	Material 1	1	HGU1 - Confining layer	Hydrostratigraphy
2	1	30	10	20	20	10	Feet above mean sea level	Material 2	1	HGU1 - Confining layer	Hydrostratigraphy
3	1	30	20	100	10	-70	Feet above mean sea level	Material 3	2	HGU2 - Aquifer	Hydrostratigraphy
4	2	20	0	10	20	10	Feet above mean sea level	Material 1	1	HGU1 - Confining layer	Hydrostratigraphy
5	2	20	10	20	10	0	Feet above mean sea level	Material 2	1	HGU1 - Confining layer	Hydrostratigraphy
6	2	20	20	100	0	-80	Feet above mean sea level	Material 3	2	HGU2 - Aquifer	Hydrostratigraphy

Record: 1 Show: All Selected Records (0 out of 6 Selected) Options

Figure 5.16. BoreholeLog table representing vertical information along two boreholes, as shown in figure 5.15.

Steps for creating 3D borehole data

This section outlines the steps involved in populating the BoreholeLog table and creating 3D BorePoint and BoreLine features. The first steps for implementing this component are to refine the geodatabase design to meet your specific project needs (see chapters 2 and 9 for a detailed description of these steps). Once the geodatabase is created, you can load vertical data into the BoreholeLog table. You can then assign WellID values to the data to associate the vertical data with Well features and translate referenced elevations (FromDepth and ToDepth) to absolute elevations (stored in the TopElev and BottomElev attributes). You can then apply tools to build 3D BorePoint and BoreLine features. Finally, you can use the results of this process to create products such as maps, scenes, and reports. The following checklist provides a summary of the main steps in creating 3D borehole data.

Checklist

1. Create the BoreholeLog table and BorePoint and BoreLine feature classes (manually or by importing from an XML schema)
2. Add project specific attributes and domains as necessary
3. Document datasets and changes made to the data model
4. Load vertical information into the BoreholeLog table
5. Assign WellID values to the vertical information, and calculate top and bottom elevations (if necessary)
6. Apply tools to create 3D features
7. Create products (maps, scenes, reports)

The following examples illustrate the process of creating 3D borehole data, including representation of completion intervals and hydrostratigraphy along boreholes. The first step is to create the BoreholeLog table and to import the vertical information into the table. Figure 5.18 shows well-completion data from the TWDB groundwater database. A selected Well feature (HydroID = 53) has seven completion intervals documented in the database. The LogType is defined as "well completion" intervals, and the elevation values represent a series of elements recorded from the land surface downward. A custom attribute, CompletionType, was added to the BoreholeLog table to distinguish between different completion types (e.g., screen, porous concrete). Similarly, appropriate attributes can be added to extend the BoreholeLog table as necessary to better represent specific datasets in your projects.

A conceptual 3D view of these data starts from the land surface and includes the completion intervals along the well. Interval depths start from 0 at the land surface and extend to 300 feet below land surface. The elevations stored in the TWDB groundwater database represent depth along the borehole measured from the land surface, in this example at 800 feet above mean sea level, as shown in the RefElev attribute. Absolute elevations are calculated by deducting the depth from the elevation of the land surface.

Another common type of vertical data is information describing subsurface strata, which is observed and classified when the well is drilled. Material, HGUID, and HGUCode attributes in the BoreholeLog table store information on subsurface formations related with intervals of the well. Material is a text attribute for storing descriptions of earth material along the borehole, and HGUID and HGUCode store hydrogeologic unit identifiers and codes to relate borehole data with defined hydrogeologic units. There is a distinction between descriptions of material, color, and texture provided in well logs and classified stratigraphy or hydrostratigraphy. Description of strata observed along the borehole is usually raw information recorded in the field that is later stored in groundwater databases. For hydrogeological analysis, the raw data are classified into conceptual stratigraphic or hydrogeologic units.

Data courtesy of Texas Water Development Board

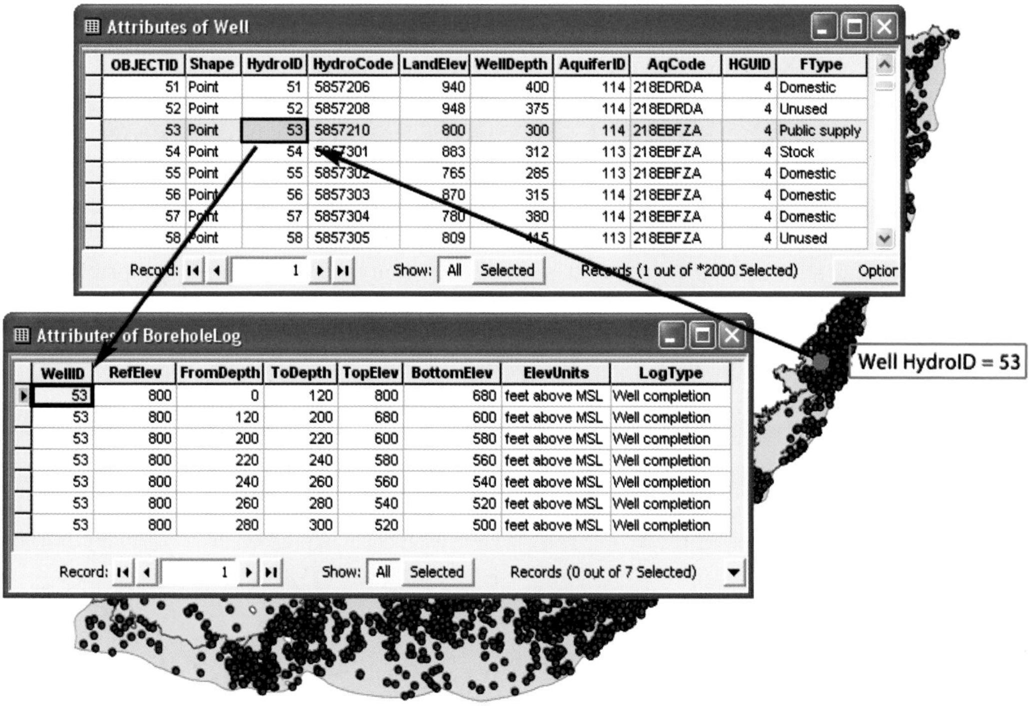

Attributes of Well

OBJECTID	Shape	HydroID	HydroCode	LandElev	WellDepth	AquiferID	AqCode	HGUID	FType
51	Point	51	5857206	940	400	114	218EDRDA	4	Domestic
52	Point	52	5857208	948	375	114	218EDRDA	4	Unused
53	Point	53	5857210	800	300	114	218EBFZA	4	Public supply
54	Point	54	5857301	883	312	113	218EBFZA	4	Stock
55	Point	55	5857302	765	285	113	218EBFZA	4	Domestic
56	Point	56	5857303	870	315	114	218EBFZA	4	Domestic
57	Point	57	5857304	780	380	114	218EBFZA	4	Domestic
58	Point	58	5857305	809	415	113	218EBFZA	4	Unused

Record: 1 Show: All Selected Records (1 out of *2000 Selected) Options

Attributes of BoreholeLog

WellID	RefElev	FromDepth	ToDepth	TopElev	BottomElev	ElevUnits	LogType
53	800	0	120	800	680	feet above MSL	Well completion
53	800	120	200	680	600	feet above MSL	Well completion
53	800	200	220	600	580	feet above MSL	Well completion
53	800	220	240	580	560	feet above MSL	Well completion
53	800	240	260	560	540	feet above MSL	Well completion
53	800	260	280	540	520	feet above MSL	Well completion
53	800	280	300	520	500	feet above MSL	Well completion

Record: 1 Show: All Selected Records (0 out of 7 Selected)

Well HydroID = 53

Figure 5.18 Example of well-construction data stored in the BoreholeLog table. The relationship between well-completion data and Well features is established by populating the WellID attribute in the BoreholeLog table.

The following example demonstrates the use of the BoreholeLog table to store hydrostratigraphic data. The USGS has developed a 3D geologic framework of the Edwards Aquifer in Bexar County (Pantea and Cole 2004). The framework includes stratigraphic data from about 450 wells, representing ten stratigraphic units and three hydrogeologic units (see a more detailed discussion on representation of 3D stratigraphy and hydrogeologic units in chapter 6). The stratigraphic data are given as a set of elevations, where each elevation represents a stratigraphic pick (the 3D location where the well penetrates a stratigraphic unit). The horizontal location is defined by x and y coordinates followed by a set of vertical (z) coordinates defining tops of formations (figure 5.19). Not all formations are represented in all wells, and the number of stratigraphic picks can range from one to the number of stratigraphic units (10 in this case).

Well features were created from the x,y coordinates, and HydroIDs were assigned to the Well features. The vertical information was imported into the BoreholeLog table, such that each row in the BoreholeLog table stores a stratigraphic pick related to a Well feature through the WellID attribute. Figure 5.20 shows an example of stratigraphy for Well 3266 (well code Ay-68-30-807).

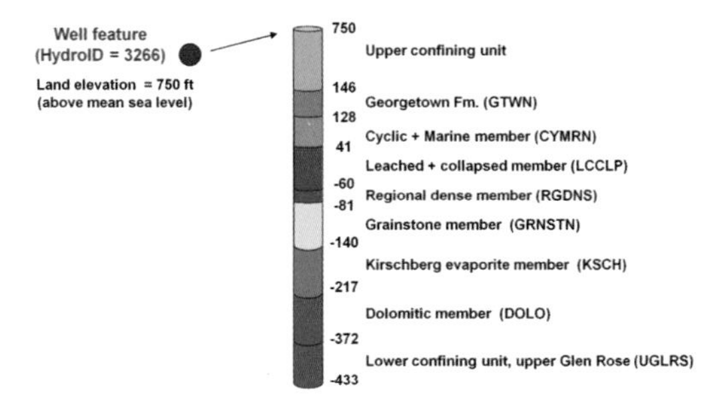

Figure 5.20 Conceptual view of 3D stratigraphic data along a borehole.

```
# Type: scattered data
# Version: 7
# Description: updated well and surface picks from USGS office San Antonio, Texas, 2/3/04
# Format: free
# Field: 1 wellid non-numeric
# Field: 2 x
# Field: 3 y
# Field: 4 LSD feet
# Field: 5 GTOWN feet
# Field: 6 CYMRN feet
# Field: 7 LCCLP feet
# Field: 8 RGDNS feet
# Field: 9 GRNSTN feet
# Field: 10 KSCH feet
# Field: 11 DOLO feet
# Field: 12 BSNOD feet
# Field: 13 UGLRS feet
# Field: 14 LGLRS feet
# Field: 15 BOTTOM feet
# Field: 16 WT feet
# Field: 17 SYMBOL integer
# Projection: Universal Transverse Mercator
# Zone: 14
# Units: meters
# Ellipsoid: Clarke 1866/NAD27
# End:
m01  562588.2127  3278912.228  ""  ""  ""  ""  950  ""  ""  ""  ""  ""  ""  ""  ""  1
m02  562664.3297  3278860.636  ""  ""  ""  ""  ""  930  ""  ""  ""  ""  ""  ""  ""  1
m03  561211.3176  3280911.12   ""  ""  ""  ""  990  ""  ""  ""  ""  ""  ""  ""  ""  1
m04  561246.599   3280998.11   ""  ""  ""  ""  ""  970  ""  ""  ""  ""  ""  ""  ""  1
m05  561409.8196  3281286.481  ""  ""  ""  ""  ""  ""  910  ""  ""  ""  ""  ""  ""  1
m06  555831.7998  3274689.727  ""  ""  ""  890  ""  ""  ""  ""  ""  ""  ""  ""  ""  1
m07  556358.8088  3276018.304  ""  ""  ""  ""  ""  890  ""  ""  ""  ""  ""  ""  ""  1
m08  556394.5713  3276020.023  ""  ""  ""  ""  910  ""  ""  ""  ""  ""  ""  ""  ""  1
m09  550489.534   3276215.977  ""  ""  ""  985  ""  ""  ""  ""  ""  ""  ""  ""  ""  1
m10  552633.0826  3275756.452  ""  ""  ""  ""  ""  900  ""  ""  ""  ""  ""  ""  ""  1
m11  552657.3674  3275797.193  ""  ""  ""  ""  920  ""  ""  ""  ""  ""  ""  ""  ""  1
m12  546867.4363  3273027.965  ""  ""  ""  875  ""  ""  ""  ""  ""  ""  ""  ""  ""  1
m13  546014.3822  3273350.38   ""  ""  955  ""  ""  ""  ""  ""  ""  ""  ""  ""  ""  1
m14  541898.3365  3270921.297  ""  ""  985  ""  ""  ""  ""  ""  ""  ""  ""  ""  ""  1
m15  543105.0051  3271446.003  ""  ""  ""  905  ""  ""  ""  ""  ""  ""  ""  ""  ""  1
m16  541060.6341  3275433.315  ""  ""  ""  ""  ""  ""  1150  ""  ""  ""  ""  ""  ""  1
m17  540937.9673  3275143.555  ""  ""  ""  ""  ""  ""  ""  1095  ""  ""  ""  ""  ""  1
m18  537630.5677  3275342.196  ""  ""  ""  ""  ""  ""  ""  1255  ""  ""  ""  ""  ""  1
m19  537905.4843  3275332.349  ""  ""  ""  ""  ""  ""  ""  ""  1125  ""  ""  ""  ""  1
```

Data courtesy of U.S. Geological Survey

Figure 5.19 Three-dimensional stratigraphic data developed as part of a 3D geologic framework of the Edwards Aquifer in Bexar County (Pantea and Cole 2004). The first column is a well identifier, followed by x and y coordinates. The next columns (4 to 16) contain a series of elevations representing stratigraphic picks (tops of stratigraphic units).

A conceptual view of this data starts from the land surface at an elevation of 750 feet above mean sea level (as recorded in the LandElev attribute of the Well feature), and a set of stratigraphic units are defined in sequence representing the stratigraphic layers down to a depth of 433 feet below mean sea level.

Figure 5.21 shows the same stratigraphic data shown in figure 5.20 stored within the BoreholeLog table. In this example, data are already given in absolute units of feet above mean sea level, thus the RefElev, FromDepth, and ToDepth attributes are not populated. The Material attribute shows the stratigraphic units, and the HGUID and

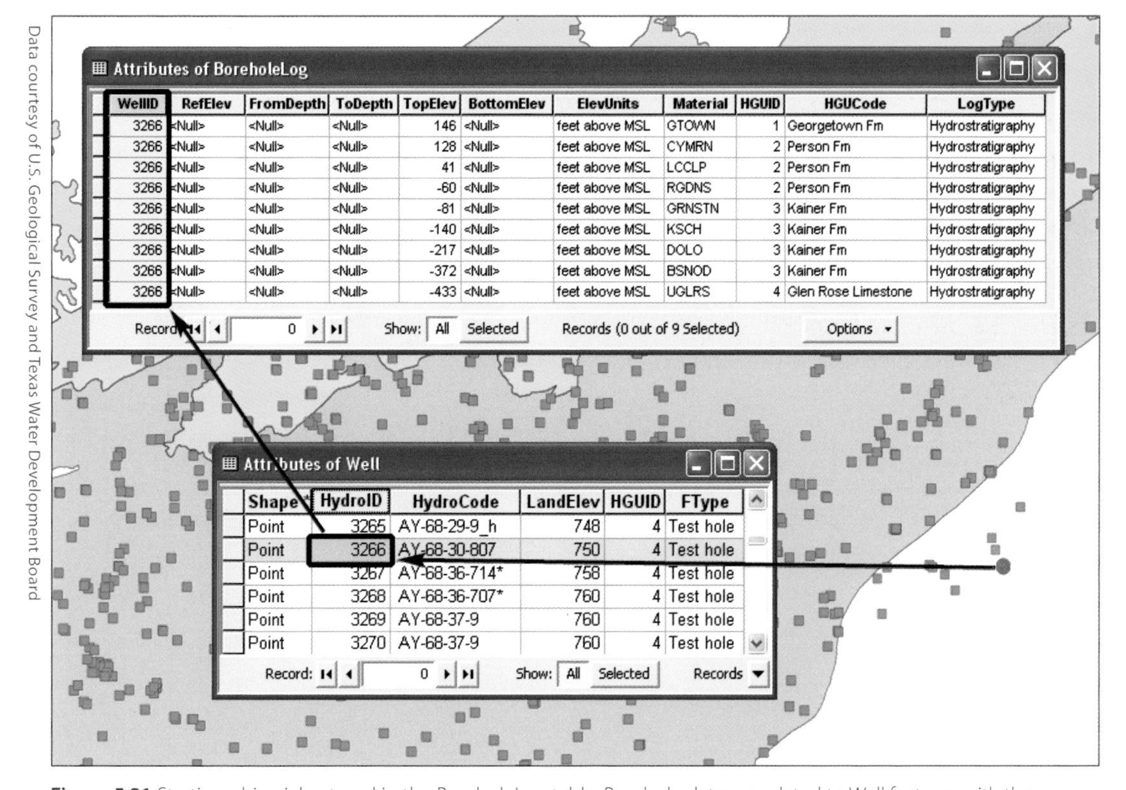

Attributes of BoreholeLog

WellID	RefElev	FromDepth	ToDepth	TopElev	BottomElev	ElevUnits	Material	HGUID	HGUCode	LogType
3266	<Null>	<Null>	<Null>	146	<Null>	feet above MSL	GTOWN	1	Georgetown Fm	Hydrostratigraphy
3266	<Null>	<Null>	<Null>	128	<Null>	feet above MSL	CYMRN	2	Person Fm	Hydrostratigraphy
3266	<Null>	<Null>	<Null>	41	<Null>	feet above MSL	LCCLP	2	Person Fm	Hydrostratigraphy
3266	<Null>	<Null>	<Null>	-60	<Null>	feet above MSL	RGDNS	2	Person Fm	Hydrostratigraphy
3266	<Null>	<Null>	<Null>	-81	<Null>	feet above MSL	GRNSTN	3	Kainer Fm	Hydrostratigraphy
3266	<Null>	<Null>	<Null>	-140	<Null>	feet above MSL	KSCH	3	Kainer Fm	Hydrostratigraphy
3266	<Null>	<Null>	<Null>	-217	<Null>	feet above MSL	DOLO	3	Kainer Fm	Hydrostratigraphy
3266	<Null>	<Null>	<Null>	-372	<Null>	feet above MSL	BSNOD	3	Kainer Fm	Hydrostratigraphy
3266	<Null>	<Null>	<Null>	-433	<Null>	feet above MSL	UGLRS	4	Glen Rose Limestone	Hydrostratigraphy

Record: 0 Show: All Selected Records (0 out of 9 Selected) Options

Attributes of Well

Shape	HydroID	HydroCode	LandElev	HGUID	FType
Point	3265	AY-68-29-9_h	748	4	Test hole
Point	3266	AY-68-30-807	750	4	Test hole
Point	3267	AY-68-36-714*	758	4	Test hole
Point	3268	AY-68-36-707*	760	4	Test hole
Point	3269	AY-68-37-9	760	4	Test hole
Point	3270	AY-68-37-9	760	4	Test hole

Record: 0 Show: All Selected Records

Data courtesy of U.S. Geological Survey and Texas Water Development Board

Figure 5.21 Stratigraphic picks stored in the BoreholeLog table. Borehole data are related to Well features with the HydroID-WellID relationship.

HGUCode show a classification of the data into hydrogeologic units.

After importing the stratigraphic data into the BoreholeLog table, custom tools can be applied to create 3D BorePoint and BoreLine features that can be visualized in ArcScene. In the following example (figure 5.22), BorePoint features were created from all the stratigraphic picks available in the dataset. The BoreLine features were created for a smaller set of wells where interval data were available (where a number of picks are available for the same well).

Vertical information recorded along boreholes is the basis for creating 3D subsurface representations. It is common to start by interpolating hydrostratigraphic data to generate views of the subsurface, such as cross sections, surfaces, and volumes. The process of creating 3D subsurface representations of hydrostratigraphy and hydrogeologic units and showing how these can be stored within the geodatabase is presented in the next chapter.

References

ASTM. 2004. D5254-92 (2004) Standard practice for minimum set of data elements to identify a groundwater site.

ANG Committee 1999. The Australian National Groundwater Data Transfer Standard Release 1.0, Australian National Groundwater Committee, Working Group on National Groundwater Data Standards.

Pantea, Michael P., and James C. Cole. 2004. Three-Dimensional Geologic Framework Modeling of Faulted Hydrostratigraphic Units within the Edwards Aquifer, Northern Bexar County, Texas. USGS Scientific Investigations Report 2004–5226.

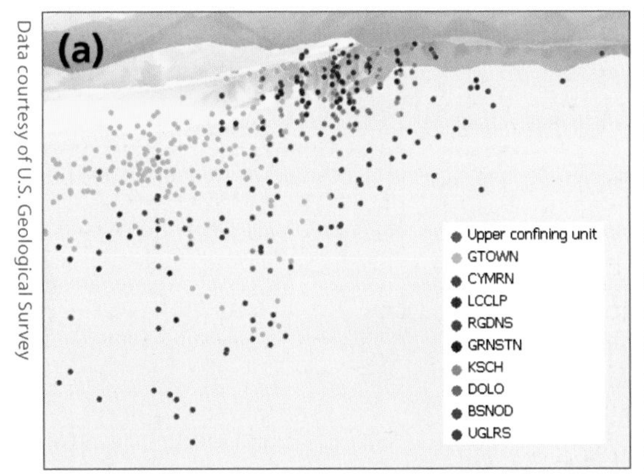

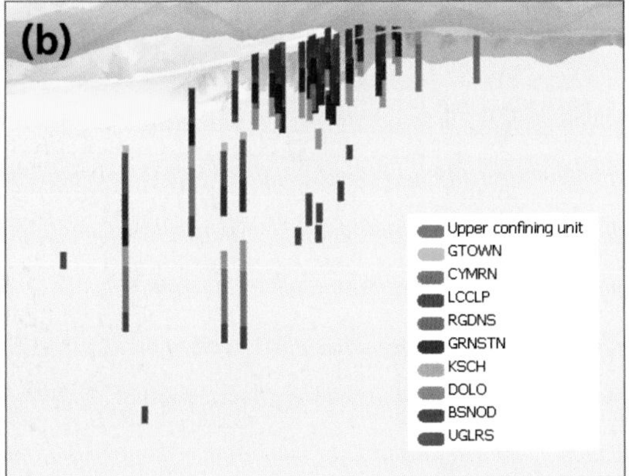

Figure 5.22 Three-dimensional view of stratigraphy displayed in ArcScene showing 3D features created from the BoreholeLog table: a) BorePoint features represent stratigraphic picks (top of formation), and b) BoreLine features represent stratigraphy as intervals.

Hydrostratigraphy

GIL STRASSBERG, NORMAN L. JONES, AND TIMOTHY WHITEAKER

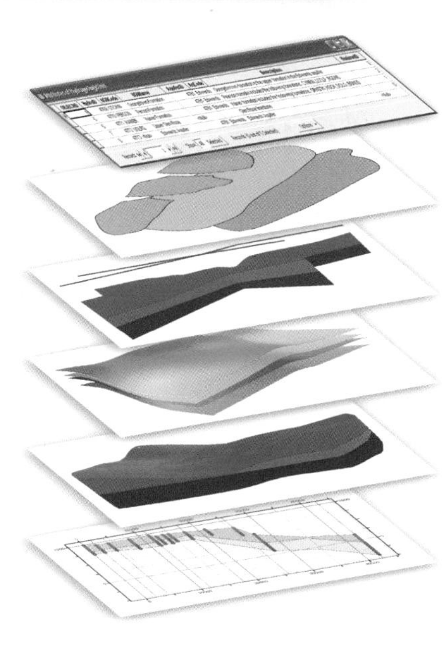

HydrogeologicUnit
Tabular representation of hydrogeologic units.
Representation: Table.

GeoArea
2D extent of hydrogeologic units.
Representation: Polygon features.

GeoSection and SectionLine
Representation of cross sections as 2D lines in plan view and as 3D features.
Representation: Polyline and Multipatch.

GeoRasters
Collection of rasters describing properties of hydrogeologic units.
Representation: Raster catalog.

GeoVolume
3D volumes representing hydrogeologic units.
Representation: Multipatch.

XS2D
2D cross section view of 3D information along a SectionLine.
Representation: Point, polyline, and polygon features displayed in a separate data frame with a unique coordinate system along a SectionLine feature

DESCRIBING THE SUBSURFACE IS A KEY PART OF MANY GROUNDWATER-RELATED projects. Because we cannot directly view and sample subsurface strata continuously, we develop conceptual models that reflect our best understanding of the arrangement and properties of subsurface strata. These conceptual models are based on observations from outcrops, boreholes, and geophysical surveys, but much of the data are inferred through interpolation and interpretation. For groundwater purposes, we develop a hydrogeologic framework, a geologic framework that defines a distinct hydrologic system. In a hydrogeologic framework, subsurface materials are classified not only by the rock properties but also by the hydraulic properties that effect water storage and flow. The creation of a hydrogeologic framework commonly involves the creation of 3D objects such as cross-sections, surfaces, and volume elements. This chapter describes the process of creating a hydrogeologic framework and how related datasets are represented in an Arc Hydro Groundwater geodatabase.

A hydrogeologic unit is any soil or rock unit or zone which by virtue of its hydraulic properties has a distinct influence on the storage or movement of groundwater (USGS glossary of hydrologic terms: `http://or.water.usgs.gov/projs_dir/willgw/glossary.html`). Establishing a hydrogeologic model that describes the spatial extent and properties of hydrogeologic units is the starting point for many groundwater projects. The terms aquifer and hydrogeologic unit are commonly used to abstract the subsurface into conceptual elements related to groundwater storage and flow. An aquifer describes a water-bearing body of subsurface strata, which in many cases is subdivided and classified for management and regulatory purposes. An aquifer or aquifer system can include a number of hydrogeologic units, which may consist of several permeable units separated by less permeable confining units. In many cases aquifers and hydrogeologic units are mapped as 2D polygons, and these mapped features are used for water-resource planning, management, and regulation. Hydrogeologic units can also be described as 3D features and visualized through a combination of mapped units, cross-sections, surfaces, and volumes. For example the Edwards Aquifer described in previous chapters was mapped as a set of

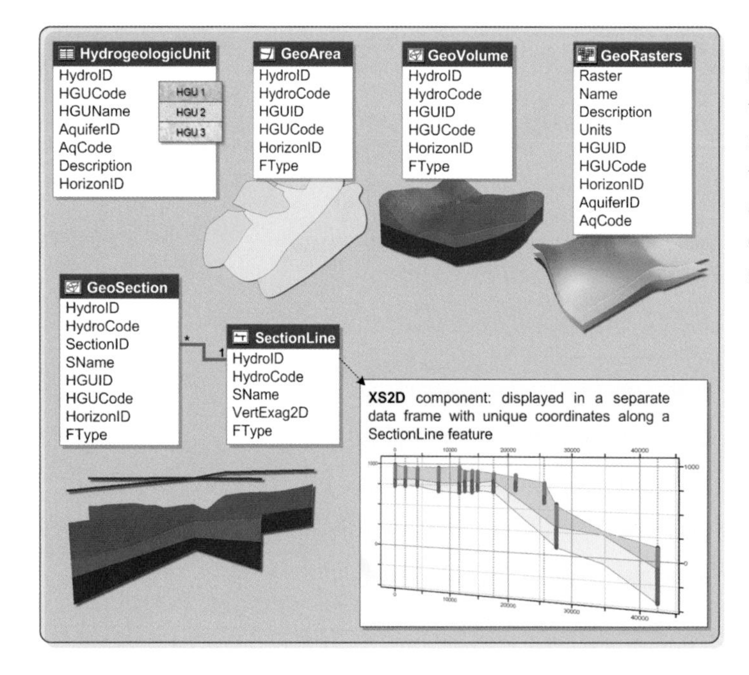

Figure 6.1 Analysis diagram showing the data elements in the hydrostratigraphy component of the groundwater data model (see figure 6.15 for a detailed description of the XS2D component representing 2D cross sections).

2D polygons defining the extent of the aquifer outcrop and confined zones. In chapter 5, a more detailed 3D dataset describes stratigraphic units of the Edwards Aquifer as 3D points and lines derived from borehole logs. The stratigraphic units described in chapter 5 can be categorized into more general hydrogeologic units used to describe groundwater flow and storage.

Geodatabase representation of hydrogeologic units

Chapter 3 described the process for creating a 3D description of hydrogeologic units. This process starts with point and interval data and continues through the creation of cross-sections and fence diagrams, the interpolation of surfaces for representing the top and bottom of hydrogeologic units, and the construction of volume objects. To support different spatial representations, the hydrostratigraphy component includes the GeoArea, GeoSection, and GeoVolume feature classes and the GeoRasters raster catalog. Each of the spatial datasets is related to a conceptual (tabular) description of hydrogeologic units stored in the HydrogeologicUnit table. An additional component for dealing with 2D cross-sections (XS2D) includes a set of feature classes representing 3D data projected on a 2D cross-section plane along a section line of interest (figure 6.1).

One of the first steps in developing a hydrogeologic model is to establish a set of conceptual units. The conceptual model may vary for different cases, depending on the scale and requirements of the project. HydrogeologicUnit is a tabular representation of hydrogeologic units. Attributes associated with the hydrogeologic units are defined in the table, and spatial features created to describe the spatial location and extent of the units are related back to the conceptual definition in the table. Attributes of the HydrogeologicUnit table are shown in figure 6.2.

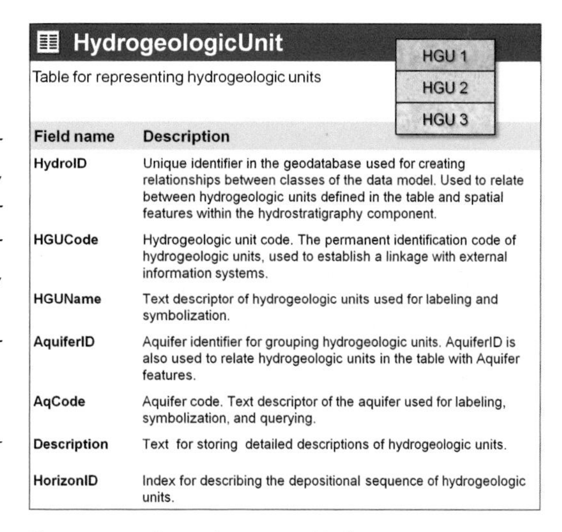

HydrogeologicUnit

Table for representing hydrogeologic units

HGU 1
HGU 2
HGU 3

Field name	Description
HydroID	Unique identifier in the geodatabase used for creating relationships between classes of the data model. Used to relate between hydrogeologic units defined in the table and spatial features within the hydrostratigraphy component.
HGUCode	Hydrogeologic unit code. The permanent identification code of hydrogeologic units, used to establish a linkage with external information systems.
HGUName	Text descriptor of hydrogeologic units used for labeling and symbolization.
AquiferID	Aquifer identifier for grouping hydrogeologic units. AquiferID is also used to relate hydrogeologic units in the table with Aquifer features.
AqCode	Aquifer code. Text descriptor of the aquifer used for labeling, symbolization, and querying.
Description	Text for storing detailed descriptions of hydrogeologic units.
HorizonID	Index for describing the depositional sequence of hydrogeologic units.

Figure 6.2 HydrogeologicUnit table for representing conceptual hydrogeologic units.

The following example shows how conceptual hydrogeologic units are grouped to define more general aquifers and confining units. For example, *Aquifer A* is a theoretical aquifer composed of three hydrogeologic units: a confining layer and two permeable formations (top and bottom units). The units are represented by three items (rows) in the HydrogeologicUnit table (figure 6.3). Each of the units is indexed with a HydroID, which will be its unique identifier within the geodatabase. The units are also indexed with the same aquifer identifier (AquiferID) and code (AqCode). In this example, the hydrogeologic units are indexed with AquiferID = 1001, which is the HydroID of an Aquifer feature that represents the boundary of *Aquifer A*. Thus, in this case three hydrogeologic units are grouped together to represent *Aquifer A* and are related to the same polygon feature representing the aquifer boundary.

Units also are indexed with a HorizonID, an index defining the vertical arrangement of hydrogeologic units in a depositional sequence. HorizonID values are assigned from bottom to top such that the smallest HorizonID is given to the base unit (the first layer in a deposition sequence), and the largest is given to the top unit. We use the HorizonID attribute to build cross-sections and volumes by filling between horizons from bottom to top (following the method by Lemon and Jones, 2003). Figure 6.4 illustrates the concept of horizons and how they relate with hydrogeologic units.

Conceptual units defined in the hydrogeologic unit table can be described by different 2D and 3D spatial datasets. The units can be represented as 2D polygon features, where each polygon defines a boundary of a unit or part of it; as 3D panels that represent cross-sections along a section line; as surfaces that define the top and bottom boundaries of units; and as volume objects that represent the 3D geometry of units. The spatial datasets all represent the same conceptual units defined in the HydrogeologicUnit table, thus each of the spatial classes is indexed also by a hydrogeologic identifier (HGUID) that relates to the conceptual description of the unit in the HydrogeologicUnit table.

GeoArea is a polygon feature class representing the 2D extent of hydrogeologic units. It is important to make a distinction between the concept of GeologyArea (presented in chapter 4) and GeoArea. GeologyArea features represent outcrops

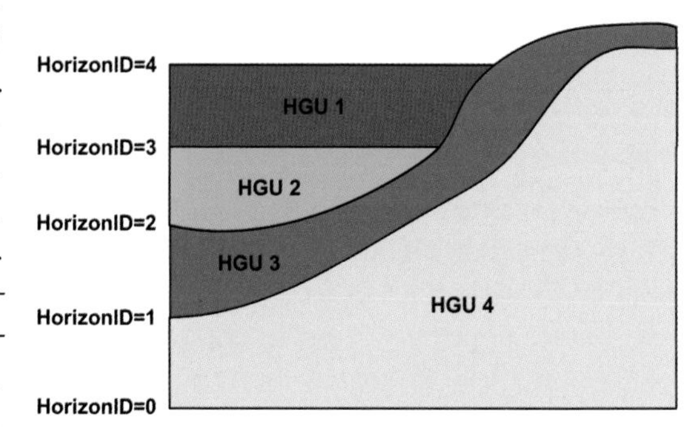

Figure 6.4 HorizonIDs define the vertical arrangement of hydrogeologic units from bottom to top.

Attributes of HydrogeologicUnit

HydroID	HGUCode	HGUName	AquiferID	AqCode	Description	HorizonID
101	HGU 1	Confining	1001	Aquifer A	Top confining layer consists of clay and sand	3
102	HGU 2	Top Unit	1001	Aquifer A	Formation consisting of permeable limestone	2
103	HGU 3	Bottom Unit	1001	Aquifer A	Formation consisting of limestone and dolomitic limestone	1
104	HGU 4	Base	<Null>	<Null>	Impermeable layer	0

Record: 0 Show: All Selected Records (0 out of 4 Selected) Options

Figure 6.3 Example of hydrogeologic units defined in the HydrogeologicUnit table.

of geologic units from surficial geologic maps, while GeoArea is more conceptual and represents boundaries of hydrogeologic units that can exist on the surface and in the subsurface. The extent of GeoArea features can be inferred and are not restricted to surficial outcrops. Attributes of GeoArea features are shown in figure 6.5.

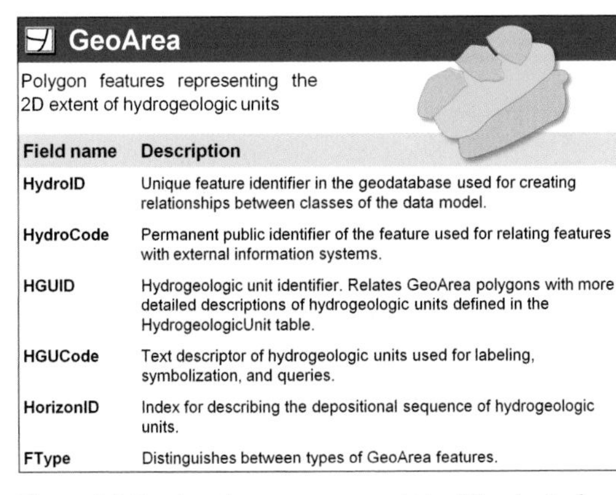

GeoArea

Polygon features representing the 2D extent of hydrogeologic units

Field name	Description
HydroID	Unique feature identifier in the geodatabase used for creating relationships between classes of the data model.
HydroCode	Permanent public identifier of the feature used for relating features with external information systems.
HGUID	Hydrogeologic unit identifier. Relates GeoArea polygons with more detailed descriptions of hydrogeologic units defined in the HydrogeologicUnit table.
HGUCode	Text descriptor of hydrogeologic units used for labeling, symbolization, and queries.
HorizonID	Index for describing the depositional sequence of hydrogeologic units.
FType	Distinguishes between types of GeoArea features.

Figure 6.5 GeoArea features represent the 2D extent of hydrogeologic units.

One hydrogeologic unit in the HydrogeologicUnit table can be associated with multiple GeoArea features. This allows mapping of discontinuous hydrogeologic units as multiple polygon features related to the same conceptual description in the HydrogeologicUnit table. Figure 6.6 shows an example of a hydrogeologic unit defined by three separate GeoArea features. In the example *HGU 1* is defined by a number of separate features, and each feature contains a unique HydroID. The features have the same HGUID and HGUCode relating to the definition of the unit in the HydrogeologicUnit table.

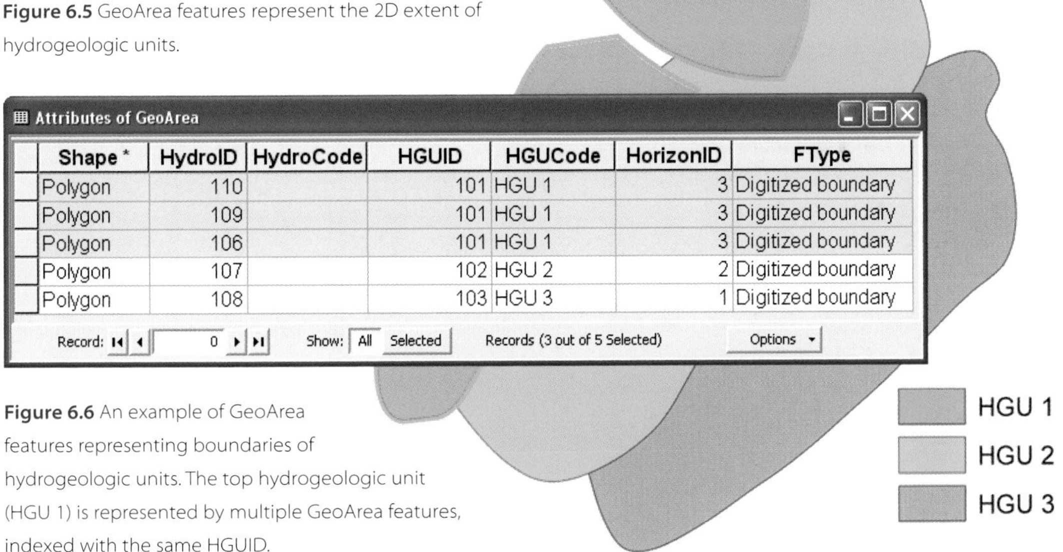

Attributes of GeoArea

Shape *	HydroID	HydroCode	HGUID	HGUCode	HorizonID	FType
Polygon	110		101	HGU 1	3	Digitized boundary
Polygon	109		101	HGU 1	3	Digitized boundary
Polygon	106		101	HGU 1	3	Digitized boundary
Polygon	107		102	HGU 2	2	Digitized boundary
Polygon	108		103	HGU 3	1	Digitized boundary

Record: 0 Show: All Selected Records (3 out of 5 Selected) Options

Figure 6.6 An example of GeoArea features representing boundaries of hydrogeologic units. The top hydrogeologic unit (HGU 1) is represented by multiple GeoArea features, indexed with the same HGUID.

Cross-sections are commonly used to illustrate the vertical dimension of hydrogeologic units along a defined section line. The first step in creating a cross-section is to define a 2D section line in plan view. Then a cross-section is created to provide a 3D view of hydrogeologic units along the section line. It is common to add additional information on the cross-section such as stratigraphy or geophysical logs recorded at boreholes along the cross-section, or water-related properties such as water levels and water quality. Figure 6.7 shows an example of a cross-section showing hydrogeologic units of the Edwards Aquifer.

We use two feature classes to represent cross-sections in the hydrostratigraphy component: SectionLine is a polyline feature class for representing 2D cross-section lines on a map, and GeoSection is a multipatch feature class representing 3D panels for constructing vertical cross-sections. Each feature in the GeoSection feature class represents a slice of a hydrogeologic unit, and a set of GeoSection features represent a complete cross-section across multiple units. Attributes of SectionLine and GeoSection features are shown in figure 6.8 and figure 6.9, respectively.

Each GeoSection feature is a panel within a 3D cross-section that represents a single hydrogeologic unit along the section line (figure 6.10). Thus one can display or query data for specific hydrogeologic units based on the HGUID and HGUCode fields. The

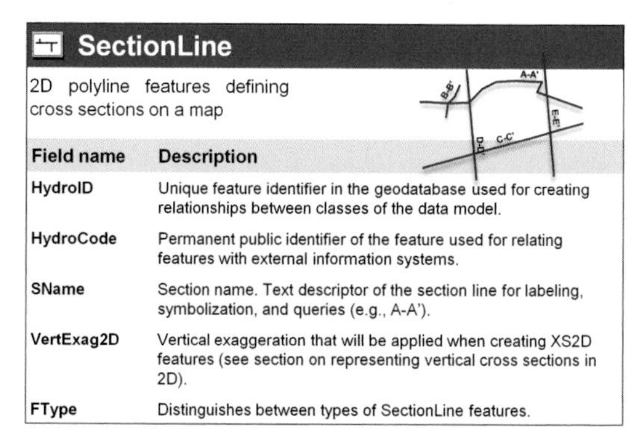

SectionLine

2D polyline features defining cross sections on a map

Field name	Description
HydroID	Unique feature identifier in the geodatabase used for creating relationships between classes of the data model.
HydroCode	Permanent public identifier of the feature used for relating features with external information systems.
SName	Section name. Text descriptor of the section line for labeling, symbolization, and queries (e.g., A-A').
VertExag2D	Vertical exaggeration that will be applied when creating XS2D features (see section on representing vertical cross sections in 2D).
FType	Distinguishes between types of SectionLine features.

Figure 6.8 SectionLine features represent cross-section lines in plan view.

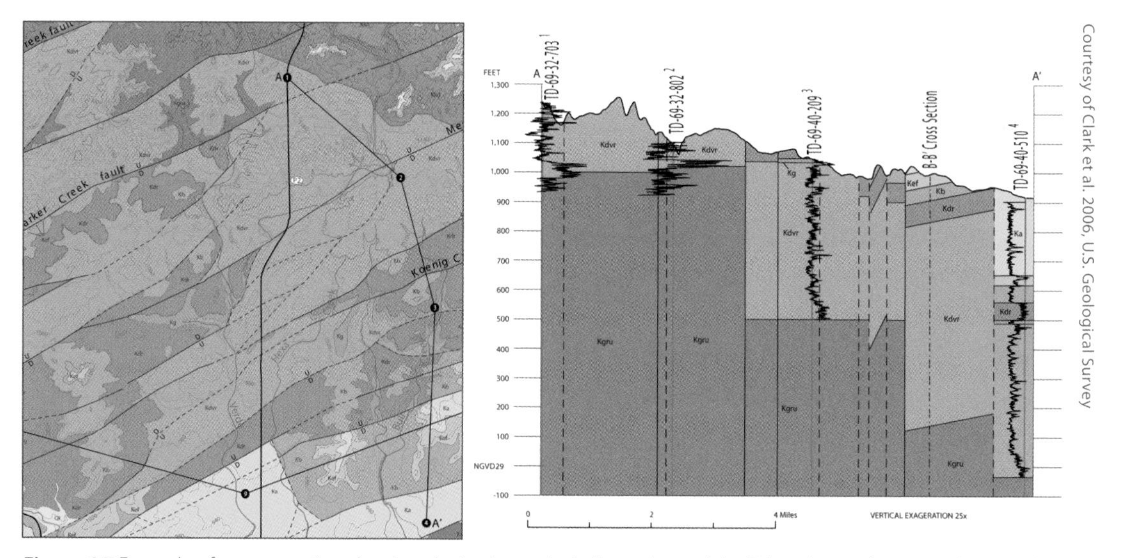

Figure 6.7 Example of a cross-section showing the hydrogeologic formations of the Edwards Aquifer in Northern Medina and Northeastern Uvalde counties, south-central Texas (figure modified from Clark et al. 2006). The left figure shows cross-section lines drawn in plan view, and the right figure shows the hydrogeologic units along section line A-A'.

relationship between SectionLine and GeoSection features is a one-to-many (1:M) relationship, where one or more GeoSection features are related with a SectionLine feature. The relationship is based on the HydroID and SectionID key fields, such that the SectionID attribute of GeoSection features is equal to the HydroID of a SectionLine feature. This enables querying a set of GeoSections along a selected section line of interest.

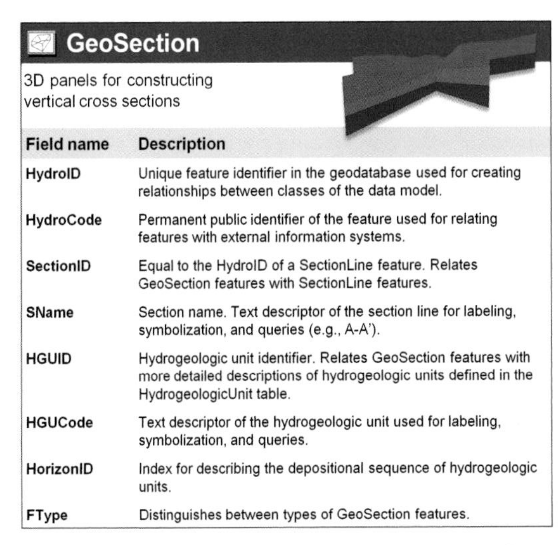

GeoSection

3D panels for constructing vertical cross sections

Field name	Description
HydroID	Unique feature identifier in the geodatabase used for creating relationships between classes of the data model.
HydroCode	Permanent public identifier of the feature used for relating features with external information systems.
SectionID	Equal to the HydroID of a SectionLine feature. Relates GeoSection features with SectionLine features.
SName	Section name. Text descriptor of the section line for labeling, symbolization, and queries (e.g., A-A').
HGUID	Hydrogeologic unit identifier. Relates GeoSection features with more detailed descriptions of hydrogeologic units defined in the HydrogeologicUnit table.
HGUCode	Text descriptor of the hydrogeologic unit used for labeling, symbolization, and queries.
HorizonID	Index for describing the depositional sequence of hydrogeologic units.
FType	Distinguishes between types of GeoSection features.

Figure 6.9 GeoSection features represent 3D panels of a cross-section.

Surfaces are commonly used for representing boundaries (top and bottom) of hydrogeologic units and for describing spatially varying hydraulic properties (e.g., hydraulic conductivity, transmissivity, and porosity) of hydrogeologic units. In ArcGIS, surfaces can be created as raster or as triangular irregular networks (TIN) datasets. Raster datasets represent imaged, sampled, or interpolated data on a uniform rectangular grid. Each cell in the raster holds a value that represents a property of the surface, such as average elevation, over the cell's area. TIN datasets represent surfaces as a triangulated network with nodes containing

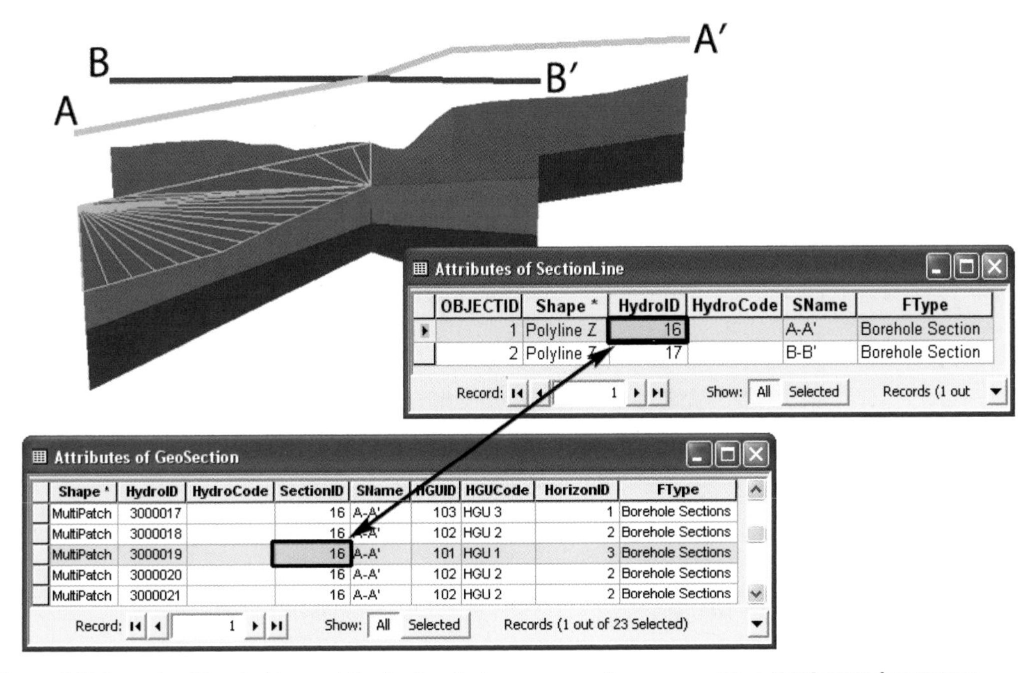

OBJECTID	Shape *	HydroID	HydroCode	SName	FType
1	Polyline Z	16		A-A'	Borehole Section
2	Polyline Z	17		B-B'	Borehole Section

Shape *	HydroID	HydroCode	SectionID	SName	HGUID	HGUCode	HorizonID	FType
MultiPatch	3000017		16	A-A'	103	HGU 3	1	Borehole Sections
MultiPatch	3000018		16	A-A'	102	HGU 2	2	Borehole Sections
MultiPatch	3000019		16	A-A'	101	HGU 1	3	Borehole Sections
MultiPatch	3000020		16	A-A'	102	HGU 2	2	Borehole Sections
MultiPatch	3000021		16	A-A'	102	HGU 2	2	Borehole Sections

Figure 6.10 Example of SectionLine and GeoSection features representing cross-sections. GeoSection features are indexed with a SectionID that is equal to the HydroID of a SectionLine feature.

surface values (e.g., elevation) and triangle edges connecting the nodes (for more information on rasters and TINs, see Zeiler 1999).

GeoRasters is a raster catalog for storing raster datasets that describe properties of hydrogeologic units. The catalog enables storing and attributing rasters within the geodatabase. For example, a set of rasters can be created for representing the top of hydrogeologic units within a 3D hydrogeologic model. Each raster dataset represents the top of a formation, and in sequence the raster surfaces define a set of 3D units, where each unit is defined by a top and bottom surface (the bottom surface is the top of the unit below). An advantage of storing raster datasets within the raster catalog is the ability to attribute the rasters. Attributes of rasters in the raster catalog can be edited within ArcMap like any other dataset. Figure 6.11 shows the attributes of the GeoRasters raster catalog.

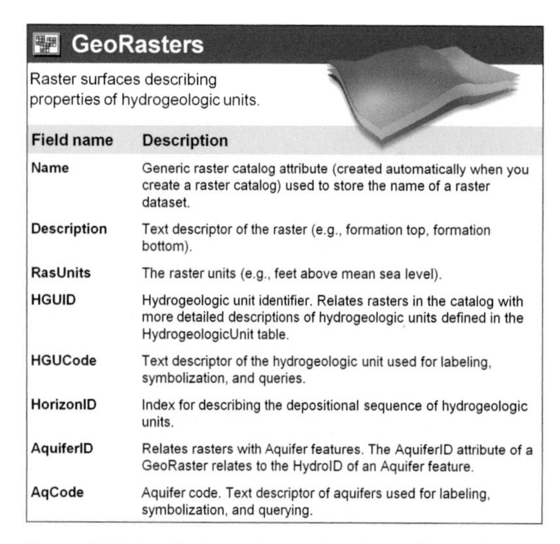

GeoRasters

Raster surfaces describing properties of hydrogeologic units.

Field name	Description
Name	Generic raster catalog attribute (created automatically when you create a raster catalog) used to store the name of a raster dataset.
Description	Text descriptor of the raster (e.g., formation top, formation bottom).
RasUnits	The raster units (e.g., feet above mean sea level).
HGUID	Hydrogeologic unit identifier. Relates rasters in the catalog with more detailed descriptions of hydrogeologic units defined in the HydrogeologicUnit table.
HGUCode	Text descriptor of the hydrogeologic unit used for labeling, symbolization, and queries.
HorizonID	Index for describing the depositional sequence of hydrogeologic units.
AquiferID	Relates rasters with Aquifer features. The AquiferID attribute of a GeoRaster relates to the HydroID of an Aquifer feature.
AqCode	Aquifer code. Text descriptor of aquifers used for labeling, symbolization, and querying.

Figure 6.11 GeoRasters raster catalog for storing and indexing raster datasets describing properties of hydrogeologic units.

The example shown in figure 6.12 shows a set of raster surfaces stored in the GeoRasters raster catalog. Each raster in the catalog represents the top of a hydrogeologic unit in feet above mean sea level. The units represented by the rasters are indexed with different HGUID values (101, 102, and 103). Three of the four rasters in the catalog are part of *Aquifer A*, thus they are indexed (in the AquiferID attribute) with the HydroID of the Aquifer feature representing *Aquifer A*, which is 1001. The rasters are also indexed with HorizonIDs to indicate the sequence of the units within the hydrogeologic framework.

GeoRasters can also represent continuous properties of aquifers or hydrogeologic units, such as conductivity, transmissivity, and specific yield

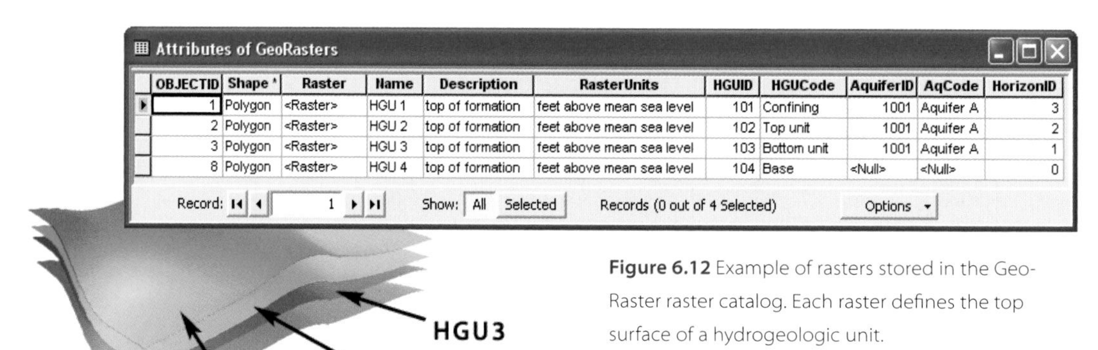

Attributes of GeoRasters

OBJECTID	Shape *	Raster	Name	Description	RasterUnits	HGUID	HGUCode	AquiferID	AqCode	HorizonID
1	Polygon	<Raster>	HGU 1	top of formation	feet above mean sea level	101	Confining	1001	Aquifer A	3
2	Polygon	<Raster>	HGU 2	top of formation	feet above mean sea level	102	Top unit	1001	Aquifer A	2
3	Polygon	<Raster>	HGU 3	top of formation	feet above mean sea level	103	Bottom unit	1001	Aquifer A	1
8	Polygon	<Raster>	HGU 4	top of formation	feet above mean sea level	104	Base	<Null>	<Null>	0

Record: 1 Show: All Selected Records (0 out of 4 Selected) Options

Figure 6.12 Example of rasters stored in the GeoRaster raster catalog. Each raster defines the top surface of a hydrogeologic unit.

that are commonly interpolated from point data. The point data can be stored as attributes of Well features, and the interpolated results can be stored as raster datasets in the GeoRasters raster catalog.

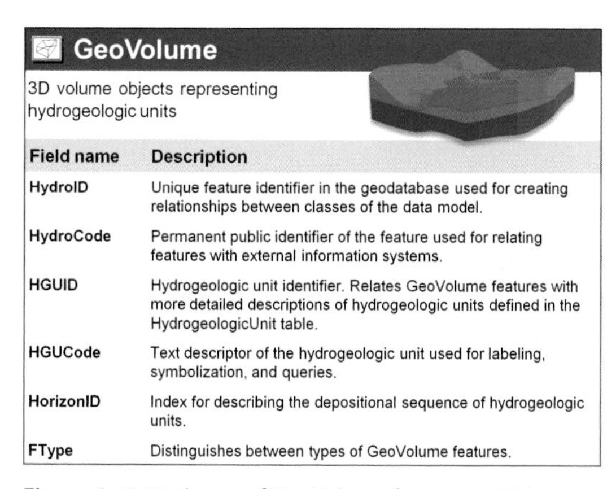

GeoVolume

3D volume objects representing hydrogeologic units

Field name	Description
HydroID	Unique feature identifier in the geodatabase used for creating relationships between classes of the data model.
HydroCode	Permanent public identifier of the feature used for relating features with external information systems.
HGUID	Hydrogeologic unit identifier. Relates GeoVolume features with more detailed descriptions of hydrogeologic units defined in the HydrogeologicUnit table.
HGUCode	Text descriptor of the hydrogeologic unit used for labeling, symbolization, and queries.
HorizonID	Index for describing the depositional sequence of hydrogeologic units.
FType	Distinguishes between types of GeoVolume features.

Figure 6.13 Attributes of GeoVolume features, which represent hydrogeologic units as 3D volumes.

Although they can be represented using different types of features (cross-sections, 3D points and lines, surfaces, etc.), hydrogeologic units are fundamentally 3D objects. When developing a 3D hydrogeologic model, it is common to develop a set of 3D volumes representing the hydrogeologic units within the model. GeoVolume is a multipatch feature class for representing hydrogeologic units as 3D volume objects (figure 6.13).

Each GeoVolume feature represents a hydrogeologic unit or a portion thereof, thus they are indexed with HGUID and HGUCode values that relate them with hydrogeologic units defined in the HydrogeologicUnit table. GeoVolume features can be displayed in ArcScene for visualizing 3D hydrogeologic models. Figure 6.14 shows an example of GeoVolume features, where each feature defines a closed volume of space representing a specific hydrogeologic unit.

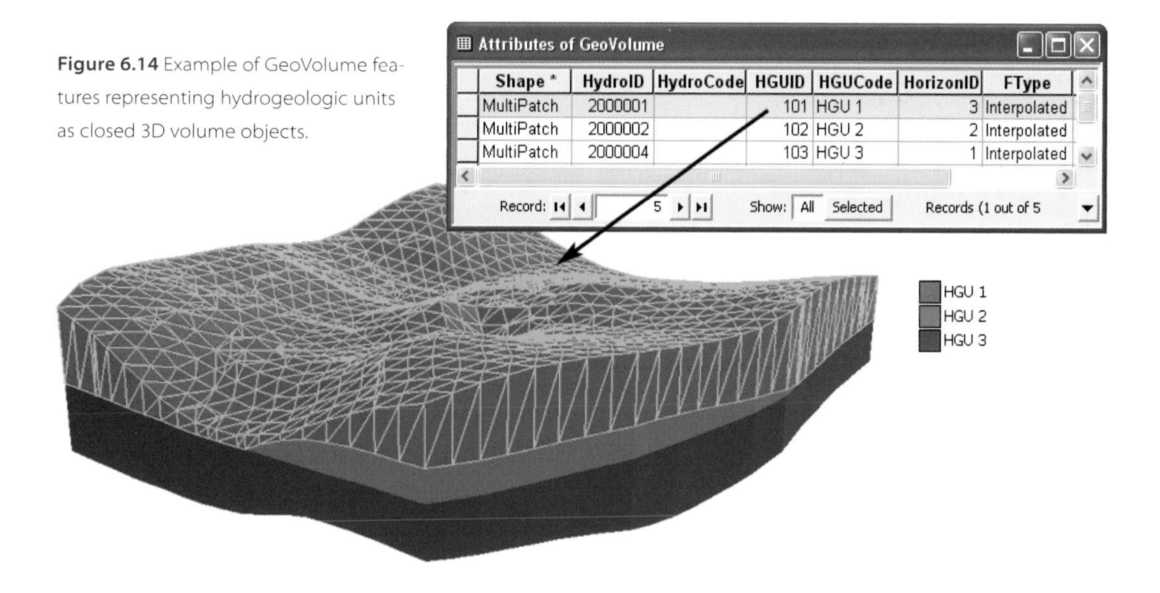

Shape *	HydroID	HydroCode	HGUID	HGUCode	HorizonID	FType
MultiPatch	2000001		101	HGU 1	3	Interpolated
MultiPatch	2000002		102	HGU 2	2	Interpolated
MultiPatch	2000004		103	HGU 3	1	Interpolated

Figure 6.14 Example of GeoVolume features representing hydrogeologic units as closed 3D volume objects.

Representing vertical cross-sections in 2D

The feature classes of the hydrostratigraphy component described thus far can be visualized both in plan and 3D view. However, when it comes to displaying vertical cross-sections taken along a section line, sometimes a special 2D view is desired. In this view the cross-section is flattened out so that on a plot of the cross-section, the horizontal axis represents the distance along the section line (which may or may not be a straight line), and the vertical axis represents depth. In other words, this 2D view translates real world (x, y, and z) coordinates into (s, z) coordinates, where *s* is the length along the section line and *z* is elevation. In addition, cross-section features are often scaled (exaggerated) in the vertical direction for better visualization.

In the groundwater data model, the 2D representation of cross-sections is implemented with multiple feature classes, and these are given the "XS2D" prefix (e.g., XS2D_Boreline and XS2D_Panel). Each XS2D feature class is associated with a single SectionLine feature, which is indexed by a HydroID. For example, if you have three section lines defined in the SectionLine feature class with HydroIDs of 1, 2, and 3, then you could have three sets of XS2D feature classes in your geodatabase, following a certain naming convention such as XS2D_Panel_1, XS2D_Panel_2, and XS2D_Panel_3. Since a single section-line feature could be associated with hundreds or even thousands of XS2D features, storing features associated with different section lines in separate XS2D feature classes helps to keep data organized. A value representing the vertical exaggeration is stored on the SectionLine features (in an attribute named VertExag2D), as it is assumed that all XS2D features associated with a given section line share the same vertical exaggeration. Figure 6.15 shows the geodatabase representation of the XS2D component.

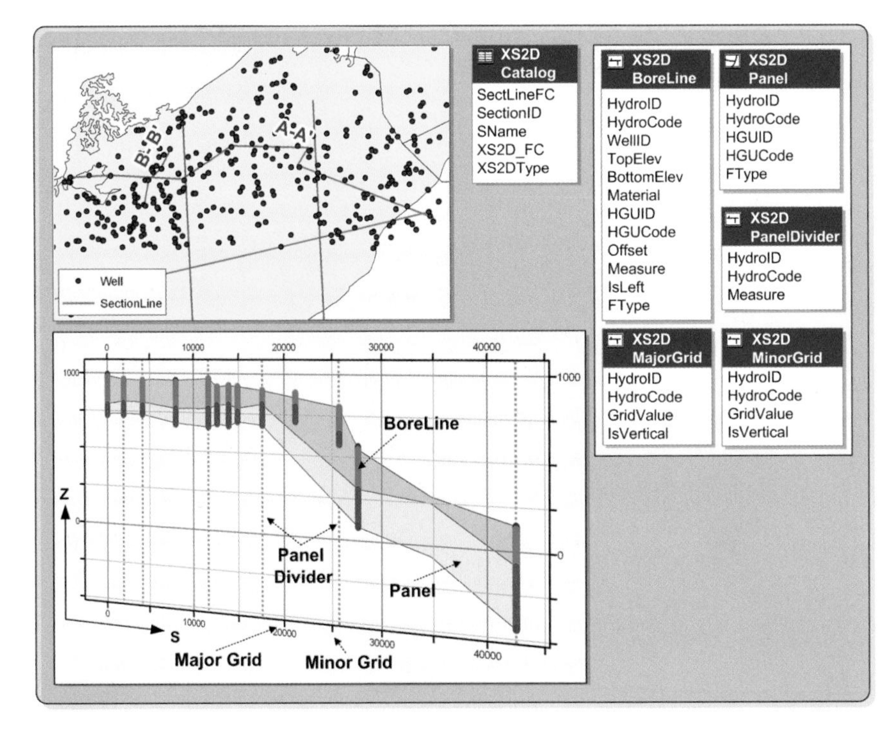

Figure 6.15 Analysis diagram showing the XS2D feature classes in the hydrostratigraphy component of the groundwater data model. The upper left figure shows SectionLine features defining the section line in plan view, and the lower figure shows a typical 2D cross-section.

XS2D_Catalog is a table for managing XS2D feature classes and their association with SectionLine features. Each row in this table provides the name and role of a XS2D feature class and information about its associated SectionLine feature (figure 6.16).

The structure of the XS2D_Catalog table allows limitless types of feature classes to be associated with a given cross-section. Common XS2D feature classes are described below, and additional feature classes can be added to represent items such as land surface elevation, water table, faults, and more.

XS2D_Panel is a polygon feature class for representing hydrogeologic units as 2D cross-section "panels." Usually a cross-section will be formed by a set of XS2D_Panel features, each representing a hydrogeologic unit along a section line. Figure 6.17 shows the attributes of the XS2D_Panel feature class.

XS2D_BoreLine is a polyline feature class for representing vertical borehole data projected on a vertical plane along a section line. Usually, when creating a new cross-section, a set of wells is selected near the line defining the cross-section, and hydrostratigraphic data along the boreholes is projected on the cross-section plane. XS2D_BoreLine features can be created from data stored in the BoreholeLog table and serve as guides for digitizing XS2D_Panel features. Figure 6.18 shows attributes of the XS2D_BoreLine feature class.

XS2D_Catalog

Table for managing XS2D feature classes and their association with SectionLine features

Field name	Description
SectLineFC	Name of the SectionLine feature class containing the referenced SectionLine feature.
SectionID	HydroID of the associated SectionLine feature.
SName	Name of the associated SectionLine feature.
XS2D_FC	Name of the XS2D feature class.
XS2DType	Role that the XS2D feature class plays in the cross section (e.g., Boreline, Panel, Grid, etc.).

Figure 6.16 XS2D_Catalog table for managing XS2D features.

XS2D_Panel

Polygon features representing hydrogeologic units as 2D cross section "panels"

Field name	Description
HydroID	Unique feature identifier in the geodatabase used for creating relationships between classes of the data model.
HydroCode	Permanent public identifier of the feature used for relating features with external information systems.
HGUID	Hydrogeologic unit identifier. Is equal to the HydroID of a hydrogeologic unit defined in the HydrogeologicUnit table.
HGUCode	Text descriptor of the hydrogeologic unit used for labeling, symbolization, and queries.
FType	Distinguishes between types of XS2D_Panel features.

Figure 6.17 XS2D_Panel features represent hydrogeologic units along a section line as 2D panels.

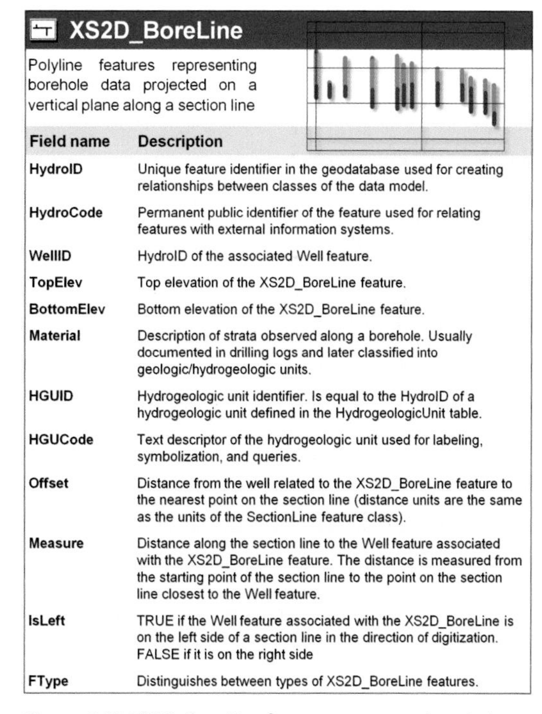

XS2D_BoreLine

Polyline features representing borehole data projected on a vertical plane along a section line

Field name	Description
HydroID	Unique feature identifier in the geodatabase used for creating relationships between classes of the data model.
HydroCode	Permanent public identifier of the feature used for relating features with external information systems.
WellID	HydroID of the associated Well feature.
TopElev	Top elevation of the XS2D_BoreLine feature.
BottomElev	Bottom elevation of the XS2D_BoreLine feature.
Material	Description of strata observed along a borehole. Usually documented in drilling logs and later classified into geologic/hydrogeologic units.
HGUID	Hydrogeologic unit identifier. Is equal to the HydroID of a hydrogeologic unit defined in the HydrogeologicUnit table.
HGUCode	Text descriptor of the hydrogeologic unit used for labeling, symbolization, and queries.
Offset	Distance from the well related to the XS2D_BoreLine feature to the nearest point on the section line (distance units are the same as the units of the SectionLine feature class).
Measure	Distance along the section line to the Well feature associated with the XS2D_BoreLine feature. The distance is measured from the starting point of the section line to the point on the section line closest to the Well feature.
IsLeft	TRUE if the Well feature associated with the XS2D_BoreLine is on the left side of a section line in the direction of digitization. FALSE if it is on the right side
FType	Distinguishes between types of XS2D_BoreLine features.

Figure 6.18 XS2D_BoreLine features represent borehole data projected on a vertical plane along a section line.

XS2D_PanelDivider is a polyline feature class representing vertical lines on a cross-section plane, showing the location where a section line changes direction (i.e., the location of vertices of the section line). Panel dividers are used as guides for orientation when viewing the 2D cross-section. Figure 6.19 shows the attributes of the XS2D_PanelDivider feature class.

XS2D_MajorGrid and XS2D_MinorGrid are polyline feature classes representing grid lines showing the vertical and horizontal dimensions in a 2D cross-section. Major grid lines are typically drawn as vertical and horizontal lines across the cross-section, while minor grid lines are indicated by tick marks along the border of the cross-section. Grid lines serve as alternative coordinate indicators in the vertical axes, because a vertical exaggeration often is applied to vertical data, and the regular ArcMap coordinate indicators do not capture the "real" coordinates of the data. Attributes of the grid line feature classes are shown in figure 6.20.

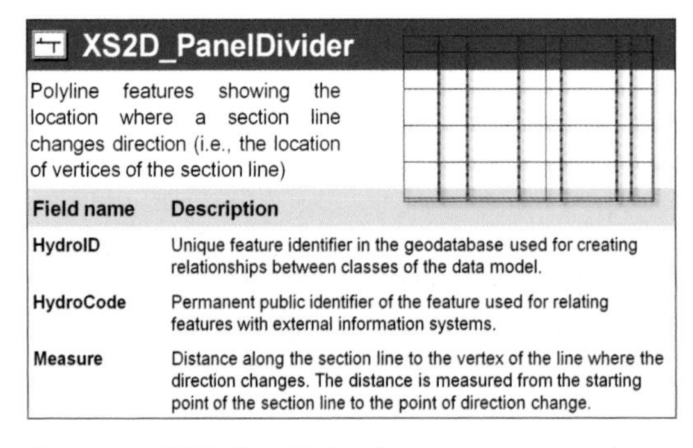

XS2D_PanelDivider

Polyline features showing the location where a section line changes direction (i.e., the location of vertices of the section line)

Field name	Description
HydroID	Unique feature identifier in the geodatabase used for creating relationships between classes of the data model.
HydroCode	Permanent public identifier of the feature used for relating features with external information systems.
Measure	Distance along the section line to the vertex of the line where the direction changes. The distance is measured from the starting point of the section line to the point of direction change.

Figure 6.19 XS2D_PanelDivider features represent points where a section line changes in direction.

Steps for populating the hydrostratigraphy component

The first steps for implementing this component are to refine the geodatabase design to meet your specific project needs (see chapters 2 and 9 for a detailed description). For example, you may want to add descriptive attributes to the HydrogeologicUnit table that provide more detailed information on units in your study. Once you have built your geodatabase, you will define hydrogeologic units in the HydrogeologicUnit table. You can then import or apply tools to create new features and assign HydroIDs to the features to create a unique identifier that will be the basis for queries and relationships within the geodatabase. Also, attribute the features with the appropriate HGUID to associate them with a hydrogeologic unit defined in the HydrogeologicUnit table. You can apply tools to create 2D cross-section views of the data using the XS2D component feature classes. Finally, you can use the results of this process to create products such as maps, scenes, and reports. The following checklist provides a summary of the main steps for creating the classes in the hydrostratigraphy component.

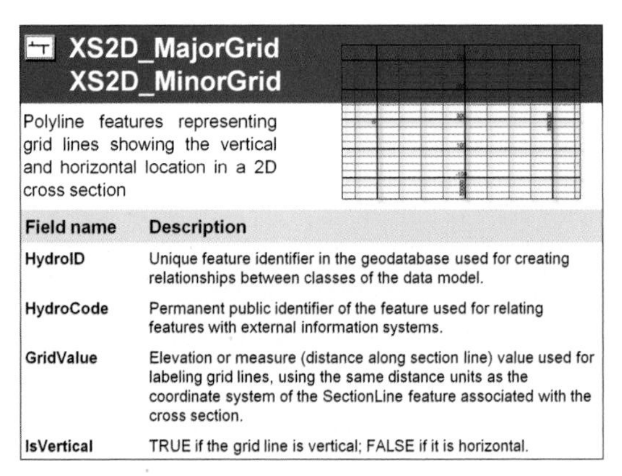

XS2D_MajorGrid
XS2D_MinorGrid

Polyline features representing grid lines showing the vertical and horizontal location in a 2D cross section

Field name	Description
HydroID	Unique feature identifier in the geodatabase used for creating relationships between classes of the data model.
HydroCode	Permanent public identifier of the feature used for relating features with external information systems.
GridValue	Elevation or measure (distance along section line) value used for labeling grid lines, using the same distance units as the coordinate system of the SectionLine feature associated with the cross section.
IsVertical	TRUE if the grid line is vertical; FALSE if it is horizontal.

Figure 6.20 XS2D_MajorGrid and XS2D_MinorGrid features represent grid lines showing the vertical and horizontal dimensions in a 2D cross-section.

Checklist

1. Create the classes of the hydrostratigraphy component (manually using ArcCatalog or by importing from an XML schema)
2. Add project specific classes, attributes, relationships, and domains as necessary
3. Document datasets and changes made to the data model
4. Define conceptual hydrogeologic units for your project in the HydrogeologicUnit table
5. Import data and apply tools to create 2D and 3D features and raster surfaces
6. Assign unique HydroIDs to features and assign HGUIDs to link them to hydrogeologic units
7. Apply tools to create 2D and 3D views of the hydrogeologic units
8. Create products (maps, scenes, reports)

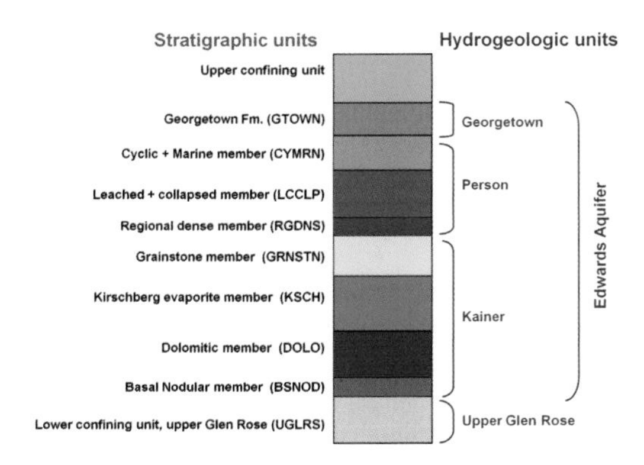

Figure 6.21 Ten stratigraphic units were classified into four hydrogeologic units. Three of the hydrogeologic units are grouped to form a more general unit, the Edwards Aquifer.

The following examples illustrate the creation of 2D and 3D features for representing the hydrogeology of the Edwards Aquifer. The first step is to define a set of hydrogeologic units for your project. Figure 6.21 shows how the ten hydrostratigraphic units described in chapter 5 were classified into four hydrogeologic units, from bottom to top: Upper Glen Rose, Kainer, Person, and Georgetown. The latter three can be grouped to form a more general hydrogeologic unit, the Edwards Aquifer.

For this example, five units were defined in the HydrogeologicUnit table (figure 6.22). Three units define the Kainer, Person, and Georgetown formations, which are part of the Edwards Aquifer, thus they are indexed with the same AquiferID (AquiferID = 4761), which references a polygon feature in the Aquifer feature class defining the general boundary of the Edwards Aquifer. The fourth hydrogeologic unit represents the Glen Rose limestone beneath the Edwards Aquifer, and the fifth hydrogeologic unit represents the Edwards Aquifer as a single hydrogeologic unit. The hydrogeologic units (except for the last one) are indexed with a HorizonID that defines the depositional sequence of the units from bottom to top. This is an important attribute for creating 3D features and keeping track of the correct 3D layering of the units.

Attributes of HydrogeologicUnit

OBJECTID	HydroID	HGUCode	HGUName	AquiferID	AqCode	Description	HorizonID
4	4769	GTOWN	Georgetown Formation	4761	Edwards	Georgetwown formation is the upper formation in the Edwards aquifer	3
5	4770	PERSON	Person Formation	4761	Edwards	Pearson formatin includes the following fomrations: CYMRN, LCCLP, RGDNS	2
6	4771	KAINER	Kainer Formation	4761	Edwards	Kainer formation includes the foolowing formations: GRNSTN, KSCH, DOLO, BSNOD	1
7	4772	UGLRS	Upper Glen Rose	<Null>		Glen Rose limestone	0
8	4773	<Null>	Edwards Aquifer	4761	Edwards	Edwards Aquifer	<Null>

Record: 1 Show: All Selected Records (0 out of 5 Selected) Options

Figure 6.22 Example of hydrogeologic units in the HydrogeologicUnit table. The Georgetown, Person, and Kainer units form the Edwards Aquifer, thus they are associated with an Aquifer feature through the AquiferID attribute.

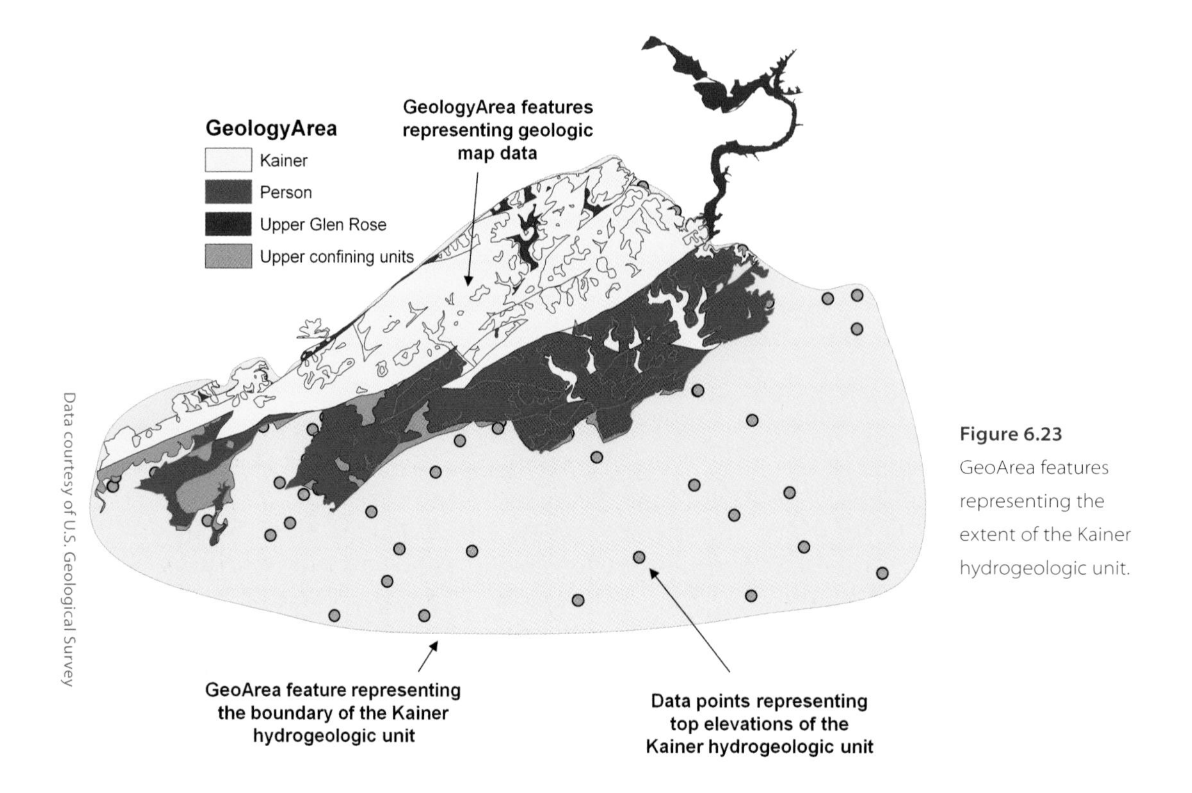

Figure 6.23
GeoArea features representing the extent of the Kainer hydrogeologic unit.

After defining conceptual units for our project, we can start populating the feature classes of the hydrostratigraphy component. Figure 6.23 shows GeoArea features representing the extent of the Kainer hydrogeologic unit that were delineated using data in the GeologyArea feature class (that define the outcrops of the hydrogeologic units), and points representing borehole hydrostratigraphy. Similarly, boundaries of other units can be delineated, and these boundaries can later serve as the basis for spatial analysis such as interpolating elevations or hydraulic properties within the defined hydrogeologic units.

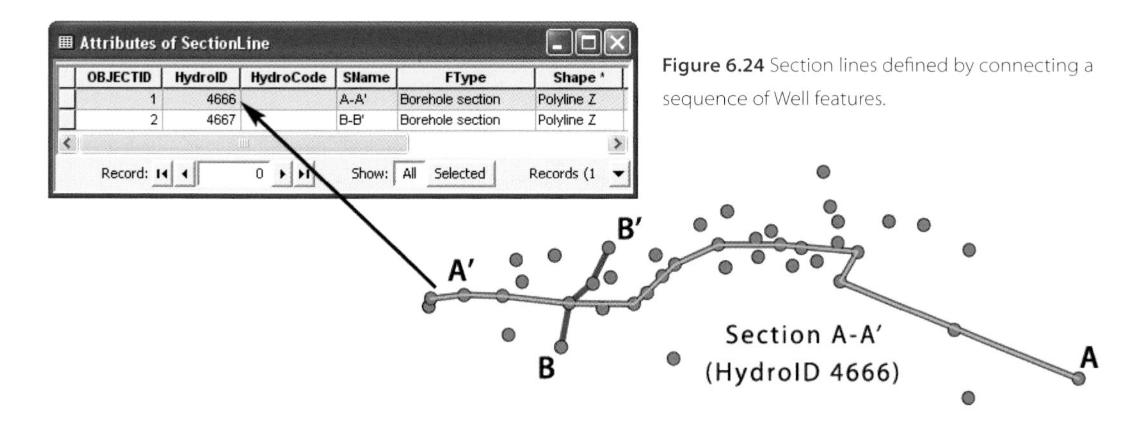

Attributes of SectionLine

OBJECTID	HydroID	HydroCode	SName	FType	Shape *
1	4666		A-A'	Borehole section	Polyline Z
2	4667		B-B'	Borehole section	Polyline Z

Figure 6.24 Section lines defined by connecting a sequence of Well features.

The next step in the process is to create cross-sections. The simplest approach for creating cross-sections is to delineate a section line connecting a sequence of wells with related 3D hydrogeologic information. As shown in figure 6.24, two section lines (A-A′ and B-B′) were drawn connecting Well features, and each of the section lines was then given a HydroID.

Once a SectionLine is defined, a 3D cross-section is created by connecting intervals of hydrogeologic units observed along the bore-holes. In the example shown in figure 6.25, Bore-Line features were connected to form a set of 3D GeoSection features. Each GeoSection in the cross-section is an individual multipatch feature that represents a single hydrogeologic unit. GeoSection

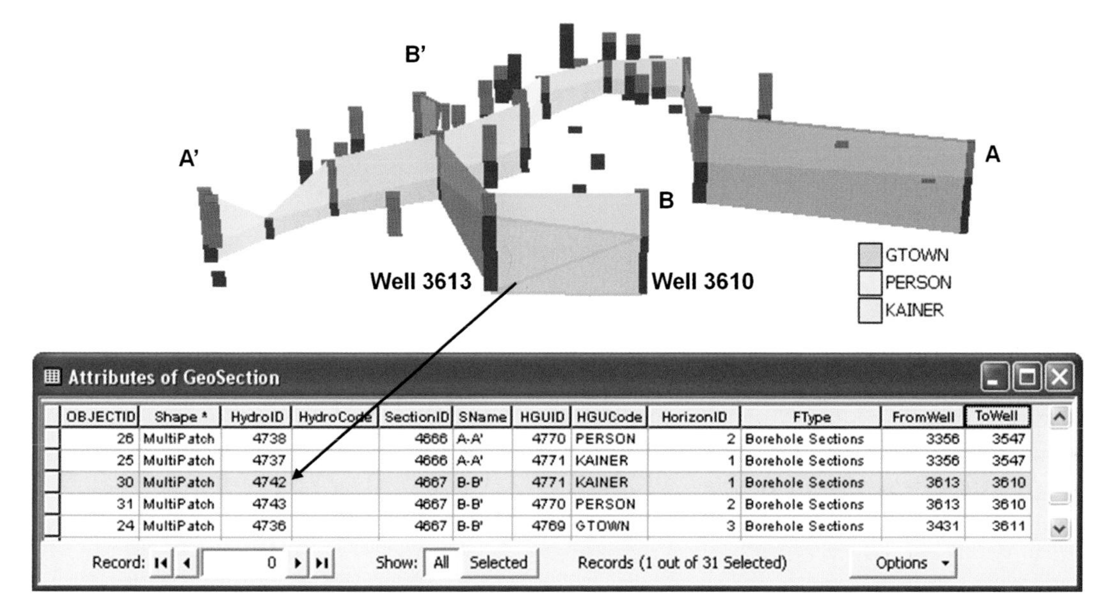

Attributes of GeoSection

OBJECTID	Shape *	HydroID	HydroCode	SectionID	SName	HGUID	HGUCode	HorizonID	FType	FromWell	ToWell
26	MultiPatch	4738		4666	A-A'	4770	PERSON	2	Borehole Sections	3356	3547
25	MultiPatch	4737		4666	A-A'	4771	KAINER	1	Borehole Sections	3356	3547
30	MultiPatch	4742		4667	B-B'	4771	KAINER	1	Borehole Sections	3613	3610
31	MultiPatch	4743		4667	B-B'	4770	PERSON	2	Borehole Sections	3613	3610
24	MultiPatch	4736		4667	B-B'	4769	GTOWN	3	Borehole Sections	3431	3611

Figure 6.25 Three-dimensional GeoSection features form a fence diagram. The GeoSections were created by connecting BoreLine features forming 3D panels representing hydrogeologic units along section lines.

features are related to a SectionLine feature via the SectionID attribute, such that we can query, display, and symbolize specific cross-sections. Two custom attributes (FromWell and ToWell) were added to store the well identifiers between which the GeoSection features were constructed. Similarly, you can extend the data model to include attributes that are useful for your specific project.

Another common approach for creating cross-sections is to develop a set of surfaces representing the top of each formation. We can then use custom tools that sample the rasters at specified intervals along the section line and construct 3D cross-sections. Figure 6.26 shows rasters representing the top of hydrogeologic units, created from hydrogeologic picks stored in the BoreholeLog table. The rasters were loaded into the GeoRasters raster catalog and indexed with a HGUID to relate them with hydrogeologic units defined in the HydrogeologicUnit table. Each of the rasters was also indexed with a HorizonID that defines the depositional sequence of the units from bottom to top. The HorizonID is used to process the rasters indexed in the GeoRasters raster catalog and derive 3D GeoSection and GeoVolume features.

In addition to representing the boundaries (top and bottom) of hydrogeologic units and aquifers, rasters can also represent physical properties of hydrogeologic units. For example, it is common to interpolate point data to create surfaces representing hydraulic properties of aquifers such as conductivity, transmissivity, and specific yield. The surfaces can then be stored and indexed within the GeoRasters catalog. This data is commonly used for groundwater calculations and analysis or fed as inputs into groundwater

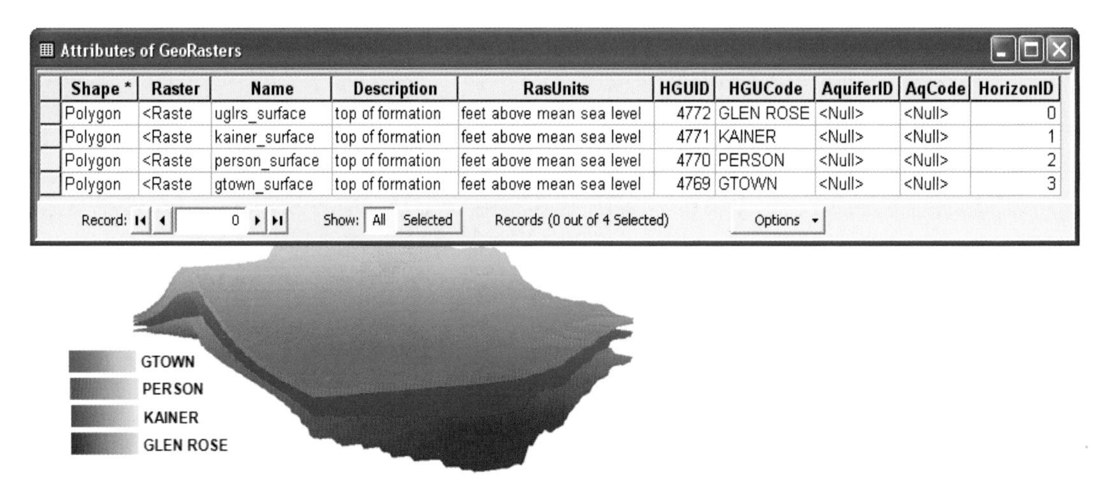

Attributes of GeoRasters

Shape *	Raster	Name	Description	RasUnits	HGUID	HGUCode	AquiferID	AqCode	HorizonID
Polygon	<Raste	uglrs_surface	top of formation	feet above mean sea level	4772	GLEN ROSE	<Null>	<Null>	0
Polygon	<Raste	kainer_surface	top of formation	feet above mean sea level	4771	KAINER	<Null>	<Null>	1
Polygon	<Raste	person_surface	top of formation	feet above mean sea level	4770	PERSON	<Null>	<Null>	2
Polygon	<Raste	gtown_surface	top of formation	feet above mean sea level	4769	GTOWN	<Null>	<Null>	3

Record: 0 Show: All Selected Records (0 out of 4 Selected) Options

Figure 6.26 Rasters representing the top of hydrogeologic units stored in a GeoRasters raster catalog. The rasters are indexed with a HGUID that relates them to definitions of hydrogeologic units in the HydrogeologicUnit table and are also indexed with a HorizonID to define the sequence of deposition from bottom to top.

Data courtesy of U.S. Geological Survey

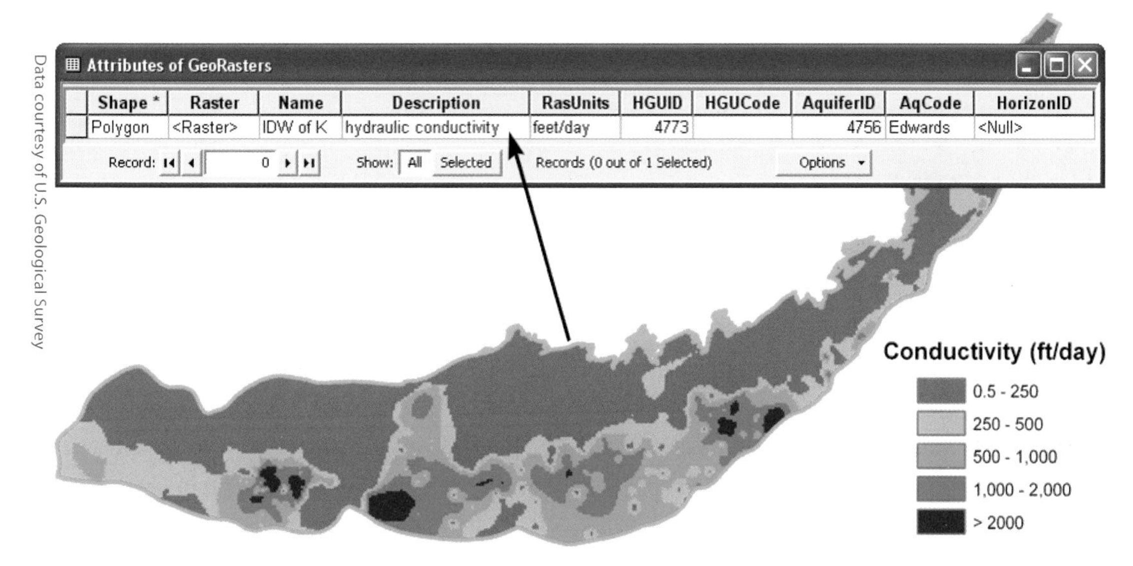

Figure 6.27 GeoRaster representing hydraulic conductivity of the Edwards Aquifer (Lindgren et al. 2004). The raster is indexed with an HGUID that relates it to the definition of a hydrogeologic unit in the HydrogeologicUnit table.

simulation models. In the following example shown in figure 6.27, a raster representing hydraulic conductivity of the Edwards Aquifer is indexed in the GeoRasters raster catalog. The raster is attributed with a HGUID to associate it with a definition of a hydrogeologic unit (defined in the HydrogeologicUnit table) and with an AquiferID to relate it with an Aquifer feature representing the boundary of the aquifer.

GeoSection features were created from the raster surfaces indexed in the GeoRasters raster catalog. Figure 6.28 shows the process of creating cross-sections: first, SectionLine features were created in ArcMap to outline the section lines of interest. Then, the Rasters To GeoSections tool (part of the Arc Hydro Groundwater Subsurface Analyst tools) was applied to derive 3D GeoSections from the raster surfaces.

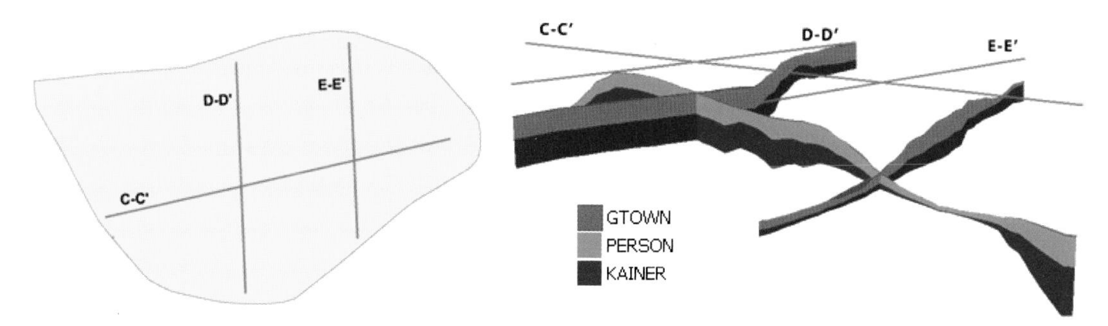

Figure 6.28 Cross-sections derived from rasters stored in the GeoRasters raster catalog. On the left, SectionLine features represent cross-sections of interest, and on the right GeoSection features derived from the raster surfaces.

GeoVolume features were created to represent hydrogeologic units as 3D volumes (figure 6.29). The volumes were built from the bottom up by filling between two consecutive raster surfaces defined in the GeoRasters raster catalog. Each GeoVolume feature is attributed with an HGUID to describe the hydrogeologic unit to which it is associated.

In addition to creating a 3D representation of the subsurface that can be visualized in ArcScene, we can create 2D cross-section views in ArcMap using the XS2D feature classes. Figure 6.30 shows the process of creating a 2D cross-section. First, a cross-section of interest is selected, and a set of wells close to the SectionLine is selected. Borehole data associated with the wells will be added to the 2D cross-section to guide the creation of panels. In addition, to better visualize the vertical data, a vertical exaggeration of 40 is defined in the VertExag2D attribute of the SectionLine feature. Using the XS2D Wizard available in Subsurface Analyst (part of the Arc Hydro Groundwater tools), a new data frame was created, and XS2D grid lines and borelines (created from hydrostratigraphy data stored in the BoreholeLog table and related with the selected Well features) were added to the XS2D data frame. Once the XS2D data frame is set, we can digitize XS2D_Panel features with standard ArcMap editing tools using XS2D_BoreLine features as guides. In addition, other data such as the land surface elevation, water levels, and well construction can be added to the cross-section view. In the following example, data from rasters of the land surface elevation and water level were added to the XS2D data frame (the process of creating a water level rasters is described in chapter 7).

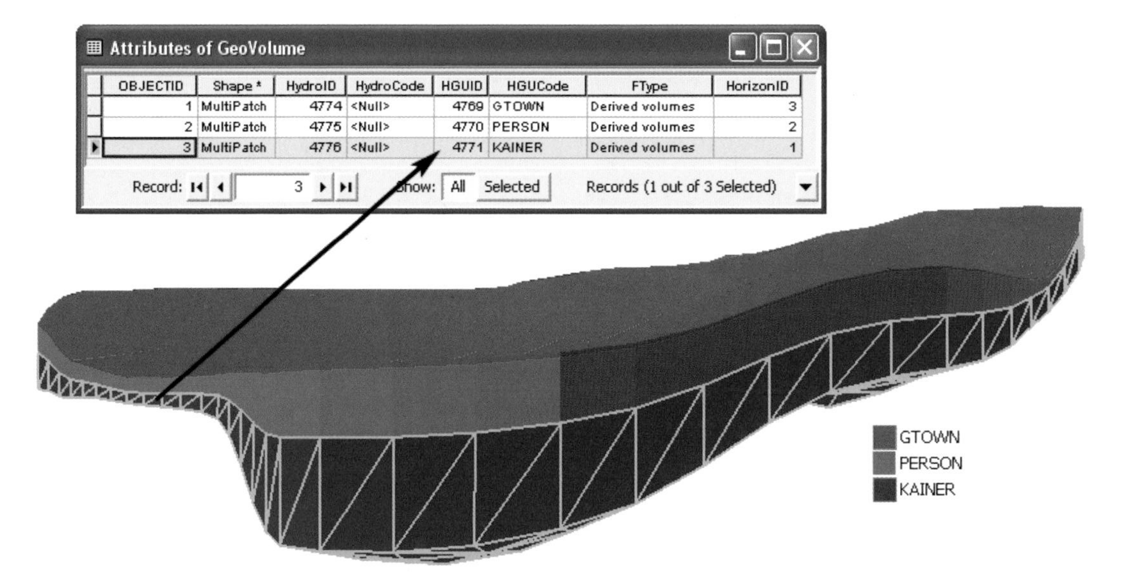

Attributes of GeoVolume

OBJECTID	Shape *	HydroID	HydroCode	HGUID	HGUCode	FType	HorizonID
1	MultiPatch	4774	<Null>	4769	GTOWN	Derived volumes	3
2	MultiPatch	4775	<Null>	4770	PERSON	Derived volumes	2
3	MultiPatch	4776	<Null>	4771	KAINER	Derived volumes	1

Record: 3 Show: All Selected Records (1 out of 3 Selected)

Figure 6.29 GeoVolume features representing 3D hydrogeologic units. Each feature represents a single volume element.

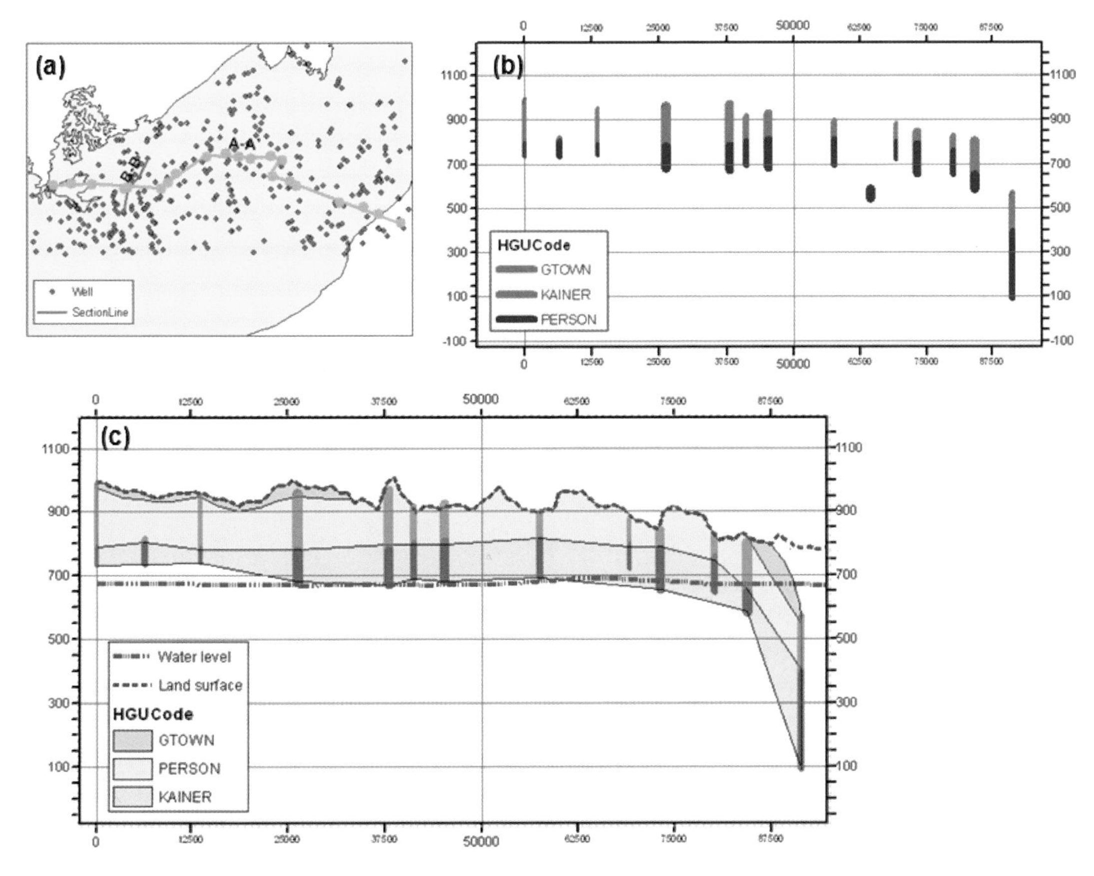

Figure 6.30 Creating a 2D cross-section with the XS2D component, a) SectionLine A-A' and nearby wells are selected; b) a new XS2D data frame is created with XS2D_BoreLine features symbolized by the hydrogeologic unit; and c) XS2D_Panels are digitized, and additional data such as the land-surface elevation and water levels are added to the cross-section view.

References

Clark, Allan K., Jason R. Faith, Charles D. Blome, and Diana E. Pedraza. 2006. Geologic Map of the Edwards Aquifer In Northern Medina and Northeastern Uvalde Counties, South-central Texas. USGS Open-File Report 2006–1372.

Lemon, A. M., and N. L. Jones. 2003. Building solid models from boreholes and user-defined cross-sections. *Computers & Geosciences* 29: 547–555.

Lindgren, R. J., A. R. Dutton, S. D. Hovorka, S. R. H. Worthington, and Scott Painter. 2004. Conceptualization and simulation of the Edwards aquifer, San Antonio region, Texas: U.S. Geological Survey Scientific Investigations Report 2004–5277, 143 p., 7 pl.

Zeiler, Michael. 1999. *Modeling Our World, The Esri Guide to Geodatabase Design*. Redlands, California: Esri Press.

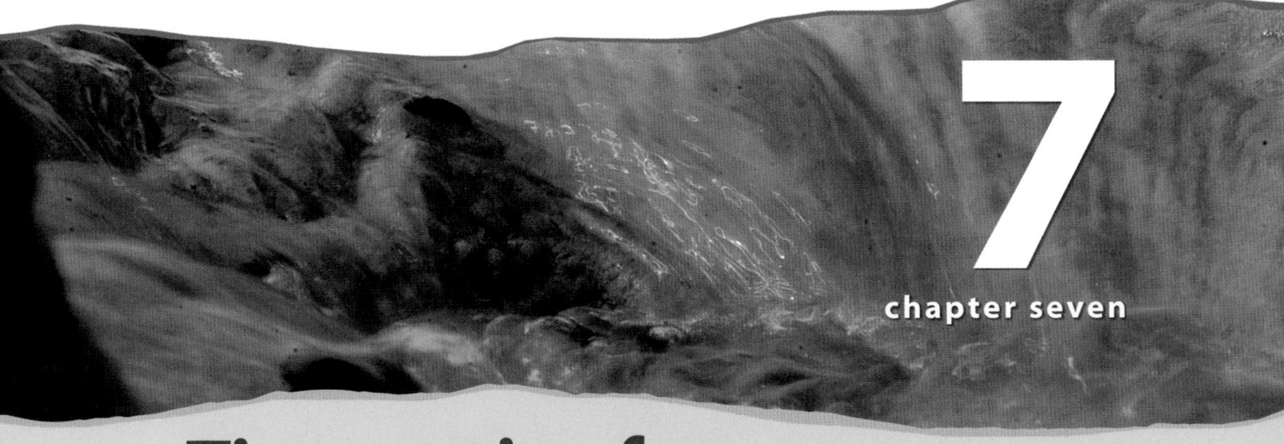

7

chapter seven

Time series for hydrologic systems

TIMOTHY WHITEAKER, DAVID R. MAIDMENT, AND GIL STRASSBERG

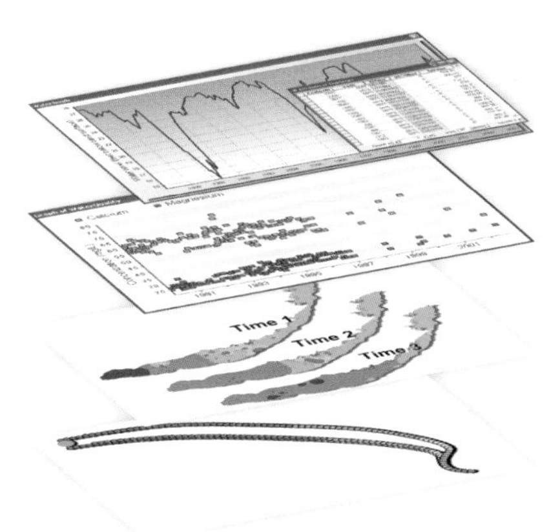

TimeSeries
Tabular description of single-variable time series, such as stream discharge and groundwater levels, where each value is recorded at a certain location in space at a given time.
Representation: Tabular date related with spatial features.

AttributeSeries
Tabular description of multi-variable time series, such as chemical concentrations in a water sample, where multiple variables are recorded at the same location and time.
Representation: Tabular date related with spatial features.

RasterSeries
Catalog for storing collection of raster datasets indexed by time.
Representation: Raster catalog.

Feature series
Collection of features indexed by time that represent a time-varying geometry (e.g. particle track, inundation polygon) varying in shape and location.
Representation: Point, line, polygon, and multipatch features indexed by time.

MUCH OF OUR UNDERSTANDING OF HYDROLOGIC SYSTEMS AND HOW WATER moves and changes in composition while it travels through them is based on measurements of water quantities and properties (e.g., flow, pressure, temperature, and concentrations) determined by measurements taken at monitoring points such as wells, gages, and sampling points along rivers. Some of the data, such as daily precipitation, daily discharge, and water levels, are collected continuously through time at fixed locations, and measurements are taken at regular intervals (e.g., every minute, hour, or day). Other data, such as water quality, are measured irregularly in time and space and might be sampled several times a year or once every couple of years. Atmospheric processes such as precipitation and evaporation play roles in the distribution of water, and these processes are often described by temporal values defined over a spatial grid.

This chapter provides a new design for dealing with temporal data within the Arc Hydro data model. The design provides better support for using multiple representations of time-series data and an improved schema for describing time-series variables. The chapter includes examples showing how to integrate temporal groundwater information within ArcGIS and also provides a more general description of implementation strategies that are useful for supporting datasets where time, location, and a set of variables are involved.

Suppose we have a geographic region, S, in which we wish to define information over a time horizon, T, for a variable, V. For groundwater, we can use the piezometric head as a simple example, measuring the water levels in an aquifer at wells and then interpolating them to form a grid or contour map representing the piezometric head in the aquifer. Because water levels do not generally vary rapidly through time, it is common when constructing piezometric head maps to choose a period of months or years, average the water levels in each well over that period, and then construct the piezometric head map by interpolating the time-averaged values. If this process is repeated for several time periods, subtracting earlier piezometric head maps from the later ones gives a measure of the change in groundwater levels within the aquifer over time.

In the example just described it is implicit that there are two types of temporal information involved: time series of water-level observations in wells and space-time datasets formed by interpreting those observations in space and time. These two concepts (time series and space-time datasets) are the cornerstones of the temporal data model presented in this chapter.

Time series

A time series is a sequence {v, t} that describes the values, v, of a variable indexed against time, t. The value of t is called the time stamp, which records an instant of time used to reference the value, v. For measuring a water level, the time stamp is the point in time at which the measurement is made. A series may be regular, that is, the time stamps (t_1, t_2, t_3, ... t_n) for successive values in the series are separated by a fixed time interval; or the series may be *irregular*, in which case there is no fixed time interval between successive values. Regular series may be produced from irregular ones—suppose the water levels measured at a well are averaged through the year for each year, y, of measurements. Then the sequence of averaged water levels indexed against the years (y_1, y_2, y_3, ...y_n) is a regular time series whose values are assumed to apply over the whole of the interval they represent. In a relational database, such as the ones used by ArcGIS, it is assumed that the time stamp for a regular time series occurs at the beginning of the time interval; for an annual time series that is midnight on the first day of the first month of each year in the series. A similar assumption is made for data averaged over any other time interval, such as daily or monthly data. A time series can thus be classified as an *instantaneous* series if the values in the series apply only at their time stamps or as an *interval* series if the values apply over the interval between one time stamp and the next.

Time-series data sources

Fortunately, many time-series datasets are now available online. Local, state, and federal agency Web sites are often a good place to look for time-series data, while universities and research labs provide alternative places for more project-specific datasets. These datasets were traditionally accessed by navigating Web pages, but more and more institutions are turning to Web services as a means of automating and standardizing data access.

Archives of groundwater data such as the U.S. Geological Survey (USGS) National Water Information System (NWIS) or the Texas Water Development Board Groundwater Database include time series that describe changes in groundwater properties over time. The most common datasets are measurements of water levels and water quality taken at wells. These data can be measured continuously at fixed locations with measurements taken at regular intervals or measured irregularly in space and time. For example, figure 7.1 shows a map of the USGS Climate Response Network, which is a national network of about 140 wells used to monitor the effects of droughts and climate variability on groundwater levels (`http://groundwaterwatch.usgs.gov`). The regularly measured water levels are summarized into statistics such as mean daily, monthly, or annual water levels, and these can be downloaded for selected wells.

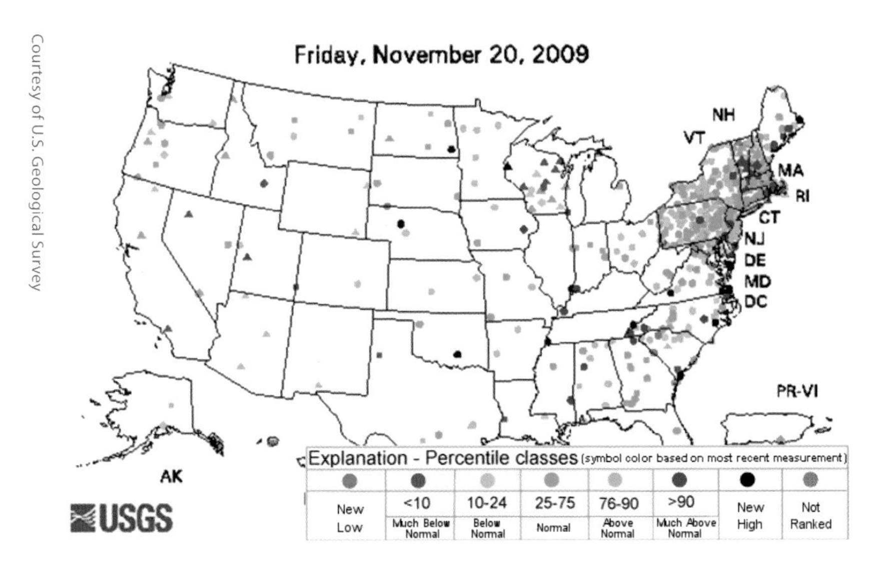

Courtesy of U.S. Geological Survey

Figure 7.1 The USGS Climate Response Network provides summary statistics of water levels showing water levels across the United States. Data are symbolized by their percentile classes so that the warmer colors indicate water levels below normal, and the cooler colors indicate water levels above normal.

Figure 7.2 shows an example of a file of daily water levels downloaded from the USGS archive. The data appear as a set of columns identifying the agency providing the data (typically "USGS"), the unique USGS site number (e.g., "295443097554201" for a well in Texas), the date and time (date/time), variable values, and data qualifiers. The type of data that a variable represents is described by a code (e.g., 01_72019_00003 = Mean depth to water level, feet below land surface), which is used as the column heading for the values. Data qualifiers provide record-level metadata that indicate, for example, if a data value has been approved for publication (qualifier code = "A").

The [time, value] pair is unique for each record, and a collection of [time, value] pairs of the same type at the same location form a time series, which can be visualized or used in hydrologic analysis and modeling. The most basic operation with time series is to visualize it in a graph that describes how a certain variable changes over time at a specific location or zone. To understand where these changes are occurring in space, we need to associate the time series with a spatial feature. For example, if we take the time series of water levels just described, we can plot the change in water level over time, and we know where these water-level changes are occurring by mapping the location of the well at which the measurements are taken. In this example, we can get the latitude and longitude coordinates for well number 295443097554201 and create a map showing the well location and the associated water-level measurements (figure 7.3).

In addition to Web pages, time-series data can also be published via Web services. A Web service is an Internet-based program built to provide a particular set of functionality or services. Whereas Web pages are accessed with Web browsers and meant to be viewed with human eyes, Web services are accessed using programming code, whose end result typically is not displayed as a Web page. A very simple example of a Web service is a function that gives you the latest quote for a stock when you provide its ticker symbol. An example of a standardized Web service for time-series data is WaterOneFlow, which was developed by the CUAHSI Hydrologic Information System (HIS) project (**http://his.cuahsi.org**).

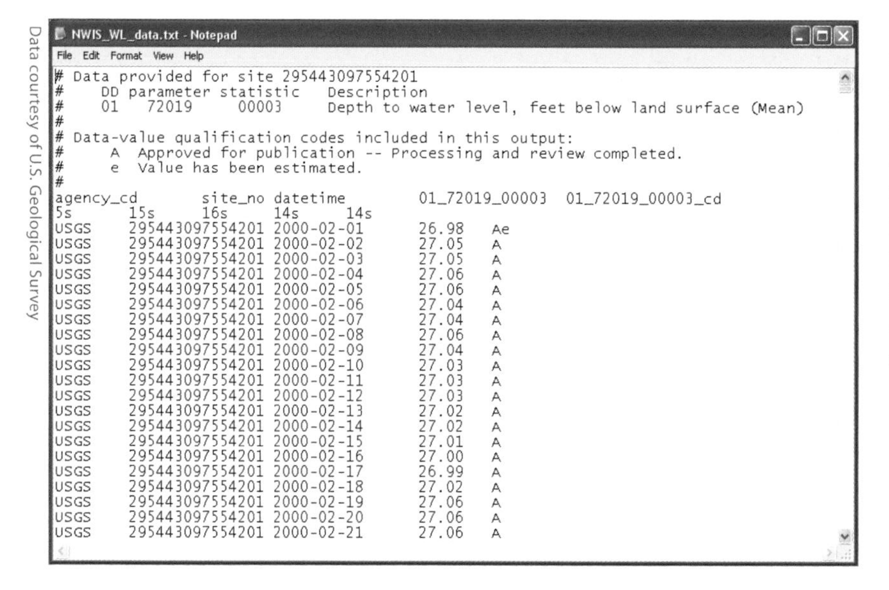
NWIS_WL_data.txt - Notepad
File Edit Format View Help

```
# Data provided for site 295443097554201
#     DD parameter statistic    Description
#     01   72019      00003     Depth to water level, feet below land surface (Mean)
#
# Data-value qualification codes included in this output:
#      A  Approved for publication -- Processing and review completed.
#      e  Value has been estimated.
#
agency_cd          site_no datetime         01_72019_00003  01_72019_00003_cd
5s         15s     16s      14s       14s
USGS       295443097554201 2000-02-01       26.98    Ae
USGS       295443097554201 2000-02-02       27.05    A
USGS       295443097554201 2000-02-03       27.05    A
USGS       295443097554201 2000-02-04       27.06    A
USGS       295443097554201 2000-02-05       27.06    A
USGS       295443097554201 2000-02-06       27.04    A
USGS       295443097554201 2000-02-07       27.04    A
USGS       295443097554201 2000-02-08       27.06    A
USGS       295443097554201 2000-02-09       27.04    A
USGS       295443097554201 2000-02-10       27.03    A
USGS       295443097554201 2000-02-11       27.03    A
USGS       295443097554201 2000-02-12       27.03    A
USGS       295443097554201 2000-02-13       27.02    A
USGS       295443097554201 2000-02-14       27.02    A
USGS       295443097554201 2000-02-15       27.01    A
USGS       295443097554201 2000-02-16       27.00    A
USGS       295443097554201 2000-02-17       26.99    A
USGS       295443097554201 2000-02-18       27.02    A
USGS       295443097554201 2000-02-19       27.06    A
USGS       295443097554201 2000-02-20       27.06    A
USGS       295443097554201 2000-02-21       27.06    A
```

Data courtesy of U.S. Geological Survey

Figure 7.2 Example of a file containing mean daily water levels downloaded from the USGS National Water Information System. Data values are indexed by site, date/time, and a code defining the variable measured.

WaterOneFlow uses a simple interface to ask for data: basically, you make one of the four following requests to retrieve data from WaterOneFlow:

- GetSites—Returns a list of observation sites (e.g., wells, stream gages) available in the Web service.
- GetSiteInfo—Returns detailed information about a particular site, such as a list of all time-series variables measured at the site and the period of record of data.
- GetVariableInfo—Returns information about a time-series variable, such as its units of measure.
- GetValues— Returns a time series of values for a given site, variable, and date range.

No matter whether the data source is the EPA, USGS, or a university, the same four basic Web service requests are made to access data, making it intuitive for a user to access data from several different sources (a list of publicly available WaterOneFlow Web services can be found at `http://hiscentral.cuahsi.org`). Additionally, WaterOneFlow Web services always return data in a standardized format called WaterML. The strict rules about naming conventions and data organization in WaterML make it easy for a data user to interpret what is returned in a given WaterML

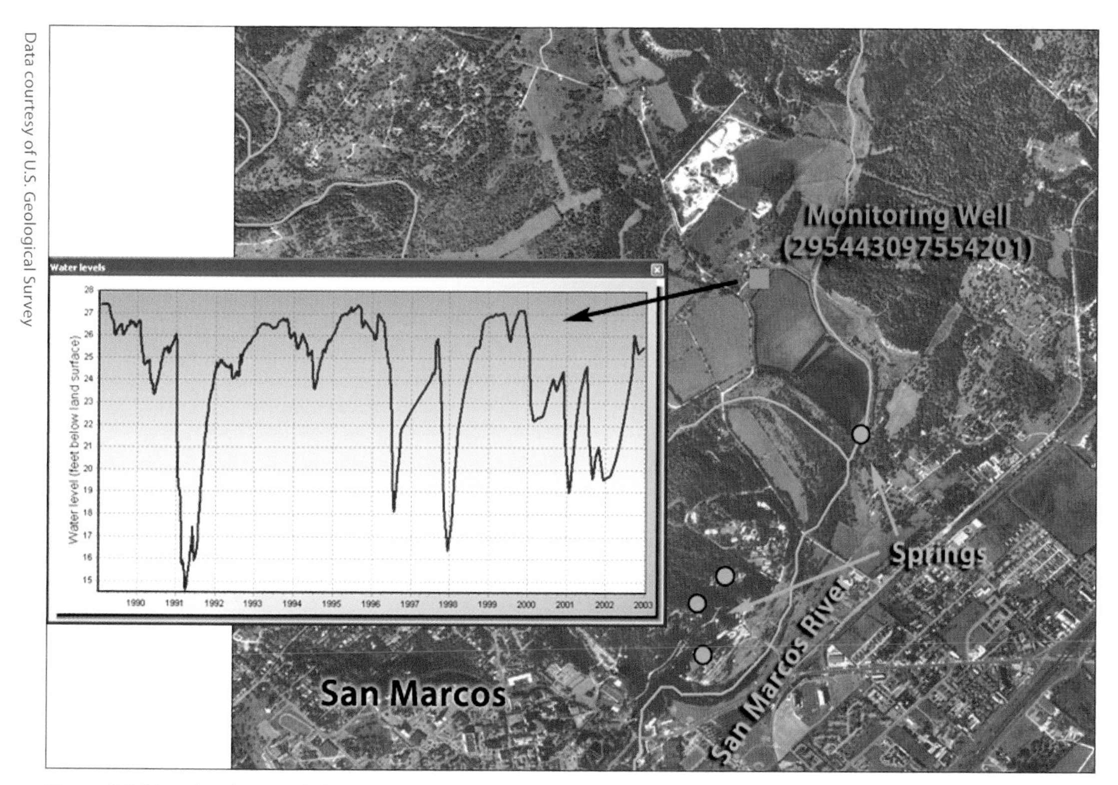

Data courtesy of U.S. Geological Survey

Figure 7.3 Water levels recorded at a monitoring well located north of San Marcos Springs in Texas. The x axis of the plot represents the time domain and the y axis shows the water level values. The well is part of the USGS NWIS real-time monitoring network.

file. Tools designed to consume the WaterML formatted Web services, such as the CUAHSI HIS HydroGET tool (`http://his.cuahsi.org/hydroget.html`) make it easy to retrieve time-series data from WaterOneFlow Web services and store the results in an Arc Hydro geodatabase. In addition, custom applications can be developed to consume Web services. An example of such an application is the NWIS Time Series Analyst developed at Utah State University. This Web-based application provides users with plotting and export functionality for data at any USGS monitoring station in the United States. Such applications, built upon Web services streamline data gathering and exploration, as they enable you to use online services instead of having to download data and use local applications for data formatting and analysis. With the adoption of Web services by more organizations, Web-based applications will be able to stream up-to-date datasets from multiple organizations through a single interface. Figure 7.4 shows an example of a plot created with NWIS Time Series Analyst. To create the plot you do not need to download data from the NWIS Web site. Instead, you simply specify the type of data you want to plot and the station number. The Time Series Analyst retrieves the data from the NWIS Web services and displays the plot and related statistics.

Another standard for data publication is Web Feature Service, or WFS. WFS is a standard way of serving vector data, defined by the Open Geospatial Consortium (`http://www.opengeospatial.org/standards/wfs`). ArcGIS Server has the capability to publish geodatabase feature classes and tables as WFS. Therefore, data providers using Arc Hydro to store time-series data can publish the temporal component online as WFS. A data user can then use ArcMap to access the tables and

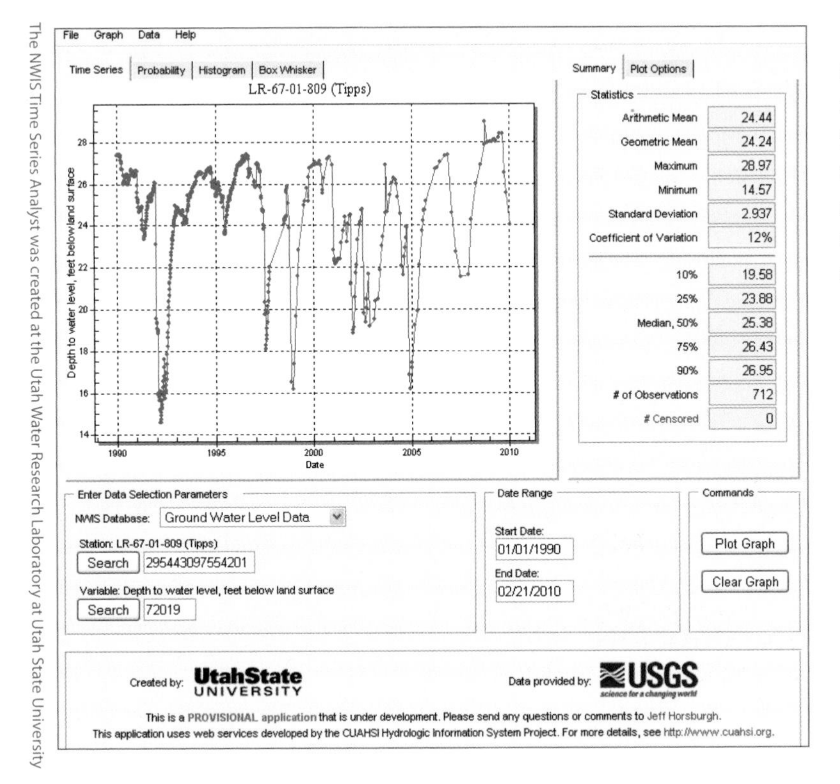

The NWIS Time Series Analyst was created at the Utah Water Research Laboratory at Utah State University

Figure 7.4 A plot of water levels created using the NWIS Time Series Analyst, a Web-services based application. The x axis shows the time dimension and the y axis shows water level values.

feature classes from the WFS so that the data appear as if they were stored in a local geodatabase on the user's computer.

By incorporating time-series data into an Arc Hydro geodatabase, we are creating a spatial-temporal information system that represents spatial features such as wells, streams, monitoring stations, and springs, and also stores time-varying information recorded at these locations. Archiving one time series of [time, value] pairs may only require a simple data structure, but in reality we might be dealing with many sets of time series, some of which can be recorded simultaneously at the same location. Moreover, for any hydrologic analysis, we will probably need to create different space-time datasets to represent the spatial and temporal distribution of hydrologic variables over time.

Space-time datasets

Time-varying data can be archived and visualized with various data structures depending on the type of data collected and the information contained in the data. Arc Hydro includes four types of space-time datasets designed to support archiving and visualizing hydrologic time series within ArcGIS:

- Time series—A single variable recorded at a location, such as stream discharge or groundwater levels.
- Attribute series—Multiple variables recorded simultaneously at the same location, such as chemical analysis of a water sample.
- Raster series—Raster datasets of a spatially continuous phenomenon indexed by time. The rasters describe how variables change spatially and temporally, such as the change in water levels in an aquifer, or the variation of precipitation measured by NEXRAD.
- Feature series—A collection of features indexed by time. Each feature in a feature series class represents a variable at a given location and time period, e.g., point features representing the movement of particles through the subsurface or polygons representing the change in inundated areas over time.

Figure 7.5 shows an analysis diagram describing the datasets included in the temporal component of Arc Hydro. The component includes the four types of space-time datasets, with additional tables for defining variables and creating summaries of time-series data.

Time-series variables

A key concept in the design of the temporal component of Arc Hydro is the definition of variables in a central table, to which different space-time datasets can be related. VariableDefinition is a table for defining temporal variables. Each variable defined in the VariableDefinition table is uniquely indexed with a HydroID, and space-time datasets are related to the variable definition by referencing the HydroID of the variable defined in the table (figure 7.6).

The following example (figure 7.7) shows a populated VariableDefinition table. Each variable in the table is indexed with a unique HydroID. The VarName, VarDesc, and VarUnits attributes describe the properties of the variable (for simplicity, not all attributes of the VariableDefinition are shown). Notice that the water-quality variables (silica, calcium, magnesium, etc.) are indexed with a VarKey. This allows us to represent these variables as attribute series, where fields in the AttributeSeries table are named based on the VarKey values in the VariableDefiniton table (see the section below on attribute series).

Single-variable time series

In a single-variable time series each value describes one variable recorded at a certain location in space at a given time. This type of data can

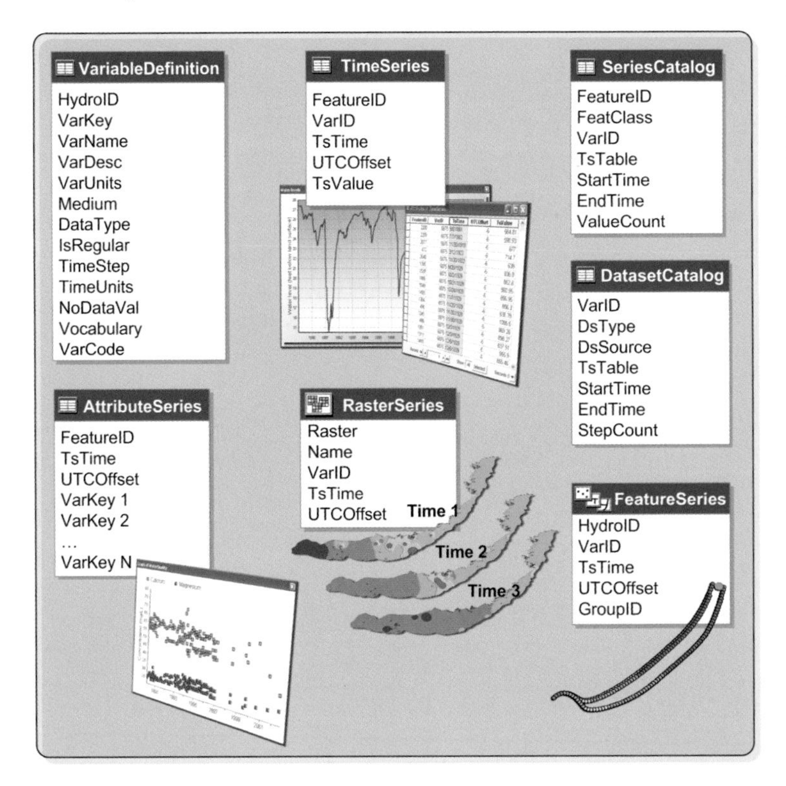

Figure 7.5 Analysis diagram showing the datasets in the temporal component of Arc Hydro.

be depicted as a 3D space-time data cube, where the three coordinates are space, indexed by location L; time indexed by T; and the variable being measured indexed by V. The data value, D, can be written as D(L,T,V) to symbolize its dependence on these coordinate axes (figure 7.8).

VariableDefinition

Table for defining temporal variables

Variable 1 – water level
Variable 2 – streamflow
Variable 3 – water quality
...
Variable N

Field name	Description
HydroID	Unique numerical identifier for the row within the geodatabase. Matches VarID in related tables.
VarKey	Unique text ID for a variable, used when a variable is indexed in an attribute series table via field names.
VarName	The name of the variable.
VarDesc	The description of the variable.
Vocabulary	Name of the list of variables in which a particular VarCode is defined (e.g., USGS NWIS).
VarCode	Public identifier for a variable (e.g., "00060" for discharge in USGS NWIS).
Medium	Medium in which the variable is observed or occurs (e.g., "Groundwater").
VarUnits	Units of measure for the variable.
DataType	Describes whether the time series contains instantaneous measurements, cumulative values, etc.
IsRegular	Integer field that stores 1 (TRUE) if the time series values are regularly spaced in time, or 0 (FALSE) if the time series is irregular.
TimeStep	For regular time series, the number of TimeUnits between each occurrence of a time series value.
TimeUnits	For regular time series, the time unit used to describe the length of time between occurrences of a time series value.
NoDataVal	Numerical value used to indicate a "No Data Value" (e.g., a missing value) in the time series table.

Figure 7.6 VariableDefinition table for defining temporal variables.

To describe time-series data values in a tabular structure, Arc Hydro defines the attributes indexing the time series. FeatureID represents the spatial feature or location (L), TsTime represents the time (T), and VarID represents the variable measured (V). Thus, any time-series value, TsValue, can be represented by a point in 3D space with its corresponding FeatureID, TsTime, and VarID (figure 7.9).

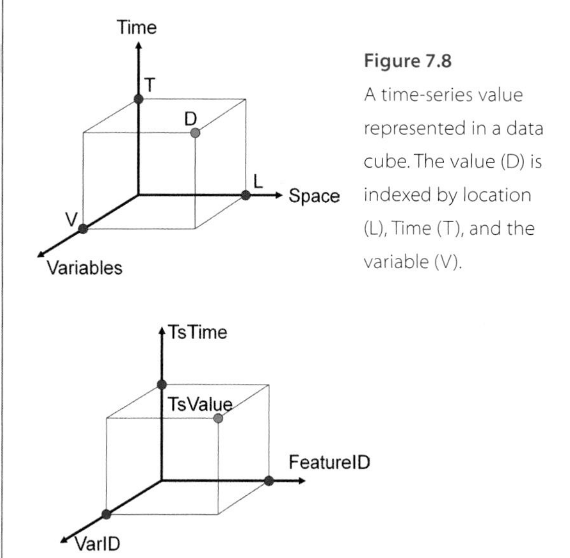

Figure 7.8 A time-series value represented in a data cube. The value (D) is indexed by location (L), Time (T), and the variable (V).

Figure 7.9 The Arc Hydro structure for describing time-series data values (TsValue). The data are indexed by the location (FeatureID), the time (TsTime), and the variable (VarID).

Attributes of VariableDefinition

OBJECTI	HydroID	VarKey	VarName	VarDesc	VarUnits
1	6874	<Null>	Streamflow	NWIS Daily Streamflow	cubic feet per second
2	6875	<Null>	Water level	Water levels	feet above mean sea level
3	6876	silica	Silica	Silica concentration as SiO2	mg\L
4	6877	calcium	Calcium	Calcium concentration as CaCO3	mg\L
5	6878	magnes	Magnesium	Magnesium concentration as Mg	mg\L
6	6879	sodium	Sodium	Sodium concentration as Na	mg\L
7	6880	potass	Potassium	Potassium concentration as K	mg\L
8	6881	AvgWL	Groundwater level	Averaged groundwater levels	feet above mean sea level
9	6882	<Null>	Particle track	Patricle track through subsurface	<Null>
10	6883	Z_Value	Depth	Depth below ground surface	feet

Record: 1 Show: All Selected Records (0 out of 10 Selected)

Figure 7.7 An example of a VariableDefinition table defining a set of time-series variables.

For data originating from different time-coordinate systems (e.g., different time zones), the UTCOffset field is attached to the TsTime field to provide an unambiguous representation of the date/time. UTCOffset is the number of hours that the TsTime is offset from Coordinated Universal Time (which is equivalent to Greenwich Mean Time). For example, in the wintertime, the Central Time Zone in the United States is six hours behind Greenwich Mean Time. So for a groundwater level value measured in Austin, Texas, at 1 p.m. local time on January 23, 2007, the TsTime is "1/23/07 1:00 PM," and the UTCOffset is "-6.0." From the viewpoint of local hydrology alone this concern with local time versus universal time may seem irrelevant because most local hydrology studies operate using the local time-coordinate system, whatever that may be, including the switch between standard time and daylight saving time. However, the treatment of universal time is relevant, even for local hydrology studies when hydrologists employ weather data such as NEXRAD radar rainfall or outputs from weather and climate models. The weather data and forecasting systems are global in extent. To avoid confusion among countries, many weather data products are presented in universal time coordinates. In order to use these data products and correctly synchronize them with local hydrologic data, the offset between local and universal time has to be understood and factored into the time-series database development.

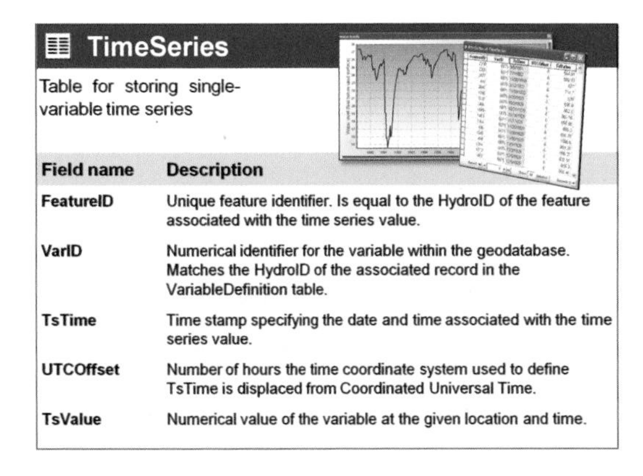

TimeSeries

Table for storing single-variable time series

Field name	Description
FeatureID	Unique feature identifier. Is equal to the HydroID of the feature associated with the time series value.
VarID	Numerical identifier for the variable within the geodatabase. Matches the HydroID of the associated record in the VariableDefinition table.
TsTime	Time stamp specifying the date and time associated with the time series value.
UTCOffset	Number of hours the time coordinate system used to define TsTime is displaced from Coordinated Universal Time.
TsValue	Numerical value of the variable at the given location and time.

Figure 7.10 TimeSeries table for representing single-variable time series.

TimeSeries is a table for storing single-variable, time-series data. The table implements the 3D structure for storing time-series values indexed by location, time, and variable. This structure is simple, yet general enough for archiving a wide array of time series. Multiple sets of time series can easily be stored using this data structure, and queries can be made on the data to extract certain variables at specific locations over a defined time interval. Each row in the TimeSeries table represents a value of a particular variable at a particular time associated with a particular feature (figure 7.10).

The TimeSeries table data structure supports the storage of many single-variable time series in

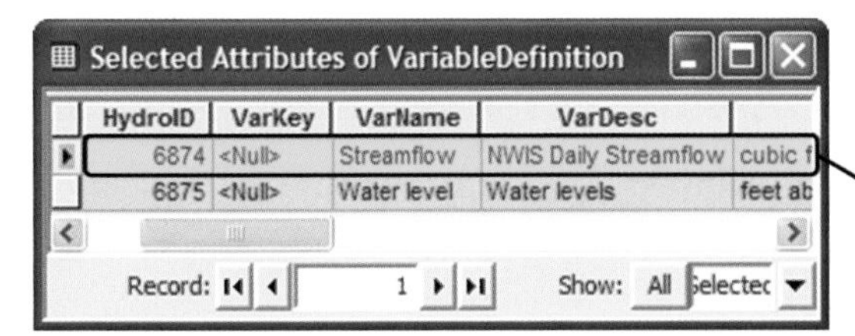

Selected Attributes of VariableDefinition

HydroID	VarKey	VarName	VarDesc	
6874	<Null>	Streamflow	NWIS Daily Streamflow	cubic f
6875	<Null>	Water level	Water levels	feet ab

Record: 1 Show: All Selectec

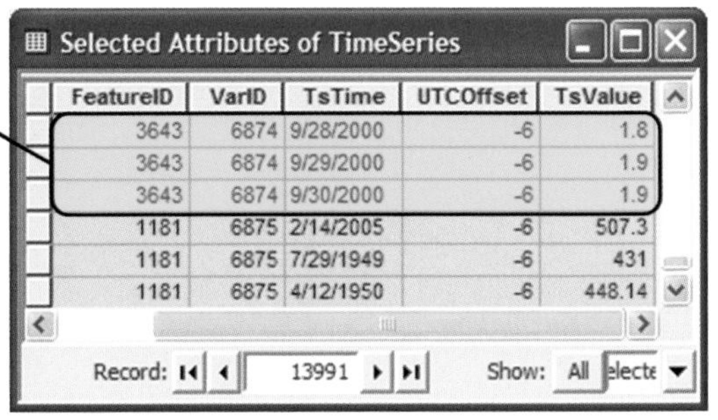

Selected Attributes of TimeSeries

FeatureID	VarID	TsTime	UTCOffset	TsValue
3643	6874	9/28/2000	-6	1.8
3643	6874	9/29/2000	-6	1.9
3643	6874	9/30/2000	-6	1.9
1181	6875	2/14/2005	-6	507.3
1181	6875	7/29/1949	-6	431
1181	6875	4/12/1950	-6	448.14

Record: 13991 Show: All Selecte

Figure 7.11 Example of time series stored in the TimeSeries table. Time-series records are indexed with a VarID that points to the HydroID of a variable defined in the VariableDefinition table.

the same table, indexed by VarID. Each row in a TimeSeries table has a VarID, which matches the HydroID of a row in the VariableDefinition table that describes the time-series variable. In the following example, shown in figure 7.11, measurements of daily stream flow downloaded from the USGS NWIS Web site for gages in the Guadalupe River basin (VarID = 6874), and groundwater level measurements from the Texas Water Development Board (VarID = 6875) are stored together in one TimeSeries table. The TimeSeries records are attributed with a VarID that associates the records with a variable defined in the VariableDefinition table.

Suppose we want to select all the time-series data measured at state well 6823302 (the same well shown in chapter 2) near Comal Springs, Texas. The HydroID of this well is 2791, thus any time-series data related to this well will be indexed with the HydroID of the Well feature. To extract time-series data related to this well, we perform a query on the TimeSeries table stating "[FeatureID] = 2791." The view created by such a query contains all the time-series data recorded at this well. This data view is represented as a vertical plane in the 3D time-series cube, perpendicular to the FeatureID axis, and can include multiple variables (e.g., water levels, water quality) measured at many TsTimes. A different query can be applied to extract all data values of a specific variable; for example, water levels. Suppose the HydroID for the water-level variable in our geodatabase is 6875. To execute our query we create a query stating "[VarID] = 6875" in the TimeSeries table. This view represents a vertical plane perpendicular to the VarID axis. Combining these two queries together, "[FeatureID] = 2791 AND [VarID] = 6875," creates a view of time series for a single variable at a single feature, represented by the vertical line in the time-series cube formed by the intersection of the two vertical planes described above. Figure 7.12 shows the different views created by querying the TimeSeries table.

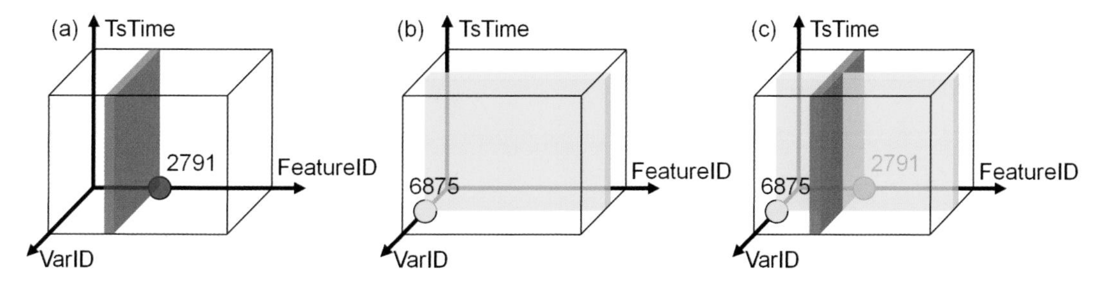

Figure 7.12 Time-series views created by queries on the TimeSeries table: (a) view of time series for a given feature (indexed by the same FeatureID), (b) view of time series of a certain variable (VarID), and (c) view of time series at a certain location and a given variable.

The dataset created from such a query has fixed FeatureID and VarID attributes and can have multiple TsTime values. The result is a single-variable time series that represents the change in water levels (VarID 6875) at the specified well (FeatureID 2791), as shown in figure 7.13.

By performing this kind of query, you can discover the various time series you have in your TimeSeries table. However, often it is useful to see a list of all available time series at a glance. For this purpose, the temporal component includes the SeriesCatalog table. SeriesCatalog is a table for indexing and summarizing time series stored in the TimeSeries table. By creating the SeriesCatalog, you can quickly identify time series of interest and then perform a query on the TimeSeries table to select the time-series values for analysis. This becomes especially useful with datasets containing a large number of time-series records because the time for processing queries may be significant. Having the summary SeriesCatalog prepopulated helps to quickly identify and extract datasets of interest (for specific variables, time periods, and features). Figure 7.14 shows the attributes of the SeriesCatalog table.

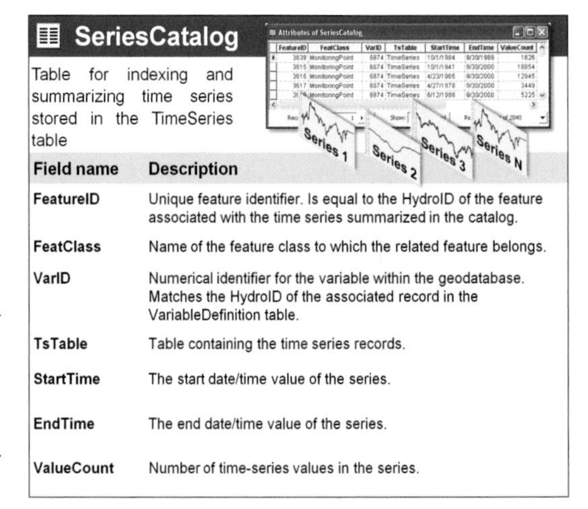

SeriesCatalog

Table for indexing and summarizing time series stored in the TimeSeries table

Field name	Description
FeatureID	Unique feature identifier. Is equal to the HydroID of the feature associated with the time series summarized in the catalog.
FeatClass	Name of the feature class to which the related feature belongs.
VarID	Numerical identifier for the variable within the geodatabase. Matches the HydroID of the associated record in the VariableDefinition table.
TsTable	Table containing the time series records.
StartTime	The start date/time value of the series.
EndTime	The end date/time value of the series.
ValueCount	Number of time-series values in the series.

Figure 7.14 SeriesCatalog table for summarizing single-variable time series stored in the TimeSeries table.

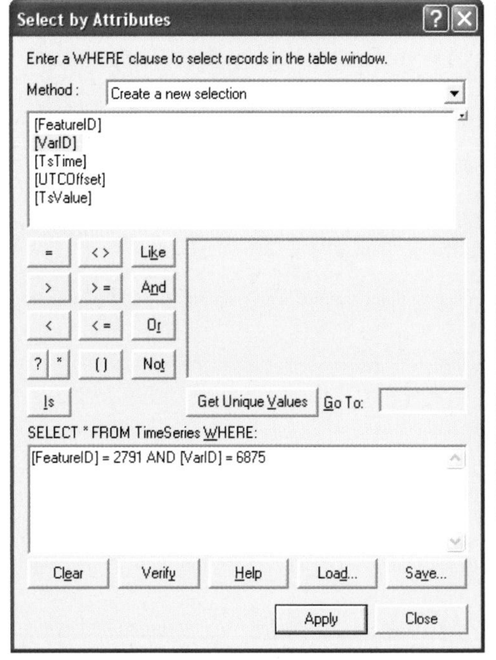

Selected Attributes of TimeSeries

FeatureID	VarID	TsTime	UTCOffset	TsValue
2791	6875	11/15/1948	-6	623.61
2791	6875	11/30/1948	-6	623.68
2791	6875	12/15/1948	-6	623.6
2791	6875	12/30/1948	-6	623.59
2791	6875	1/15/1949	-6	623.51
2791	6875	1/30/1949	-6	623.54
2791	6875	2/15/1949	-6	623.48
2791	6875	2/28/1950	-6	624.44
2791	6875	3/15/1950	-6	624.48
2791	6875	3/30/1950	-6	624.17
2791	6875	4/15/1950	-6	624.05
2791	6875	4/30/1950	-6	624.15
2791	6875	5/15/1950	-6	624.14
2791	6875	5/30/1950	-6	624.13
2791	6875	6/15/1950	-6	624.14
2791	6875	6/30/1950	-6	623.85
2791	6875	7/15/1950	-6	623.79

Record: 0 Show: All Selected Records

Figure 7.13 An example query on the TimeSeries table that creates a view of a single-variable time series at a specific feature.

The following example (figure 7.15) shows a populated SeriesCatalog table. The highlighted series in the catalog is indexed with a FeatureID of 2791 and a FeatClass value of "Well," meaning that the values recorded in this series are related to a Well feature with HydroID = 2791. The VarID value of 6875 relates to the water-level variable defined in the VariableDefinition table. The ValueCount attribute indicates that 1167 water level values were recorded at this well, between November 15, 1948, and August 27, 2003.

Customizations can be made to the SeriesCatalog table if desired. For example, a GroupName attribute could be added as a means of grouping individual time series into categories such as "groundwater" or "surface water." Sometimes, a given time series applies to more than one geospatial feature. For example, data measured at a rainfall gage could be applied to a watershed and also a county. In this case, a HydroID field is added to the SeriesCatalog to uniquely identify each series. The HydroID of the series can be used in a separate table that pairs SeriesIDs and FeatureIDs. This table essentially creates a many-to-many association between time series and features, such that a series can be associated with multiple features, and a feature can be associated with multiple series. Within the SeriesCatalog table, the FeatureID is still used, storing the HydroID of the feature where the time series was actually measured, e.g., the rainfall gage.

Until this point we have demonstrated the management of single-variable time series and the ability to query a certain variable at a given location over a specified time. Other common tasks performed with GIS are mapping time series for a given time across multiple features and the animation of variables over space and time. In the space-time cube, the combination of a horizontal slice for a particular TsTime and a particular VarID results in a view containing all data of a certain type (e.g., groundwater levels) recorded at the specified time. This data includes measurements at multiple features. It is common to create maps representing the spatial distribution of the variable at the specified time. Examples are the creation of maps that show the spatial distribution of groundwater

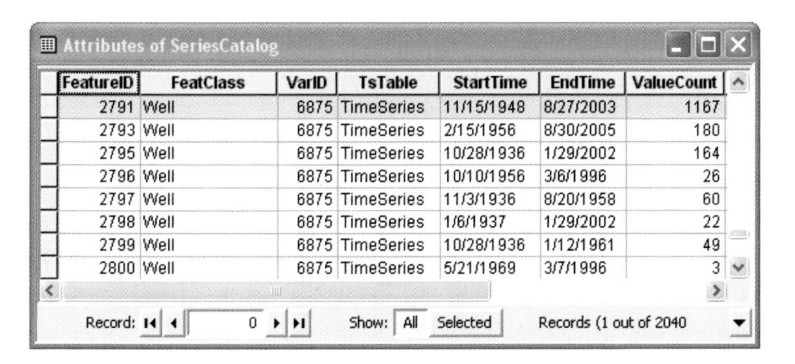
Attributes of SeriesCatalog

FeatureID	FeatClass	VarID	TsTable	StartTime	EndTime	ValueCount
2791	Well	6875	TimeSeries	11/15/1948	8/27/2003	1167
2793	Well	6875	TimeSeries	2/15/1956	8/30/2005	180
2795	Well	6875	TimeSeries	10/28/1936	1/29/2002	164
2796	Well	6875	TimeSeries	10/10/1956	3/6/1996	26
2797	Well	6875	TimeSeries	11/3/1936	8/20/1958	60
2798	Well	6875	TimeSeries	1/6/1937	1/29/2002	22
2799	Well	6875	TimeSeries	10/28/1936	1/12/1961	49
2800	Well	6875	TimeSeries	5/21/1969	3/7/1996	3

Record: 0 Show: All Selected Records (1 out of 2040

Figure 7.15 An example of a populated SeriesCatalog table showing summaries of time series.

levels and water quality within an aquifer. The following query (figure 7.16) on the time-series table retrieves all the groundwater levels measured in January 1991.

The result is a view of measurements taken at wells within the Edwards Aquifer for the time period specified. By joining the queried time series with Well features, we can create a map of water levels within the aquifer for the selected date. Using the animation tools built within ArcGIS, we can animate a set of maps showing the spatial-temporal variations in water levels within the Edwards Aquifer. The water-level maps in figure 7.17 were created by querying the Time-Series table for a specific variable and date, and the results were joined with the Well feature class.

The examples above demonstrate a conceptual process by which time-series data are visualized. Time series are related to features that locate them in space, and by using the relationship between features and time series, we can plot time series for particular features or create maps showing the spatial distribution of a variable at a given time. The TimeSeries table allows easy storage of time-series data within a simple table structure. Because the TimeSeries table has only one value attribute (TsValue), it is most useful for storing single-variable time series where one variable is recorded for any location at a specific time. Some datasets are not suitable for such a table because they actually contain multiple variables for a specific location recorded at the same time. Common examples are datasets describing chemical analyses where multiple variables are analyzed from a single water sample taken at the same location and time. Another example is the inputs and outputs of simulation models, where it is common to have multiple variables (e.g., pressure, velocity, storage) calculated for every model time step.

Attribute series

AttributeSeries is a table for archiving time-series data where multiple variables are indexed with

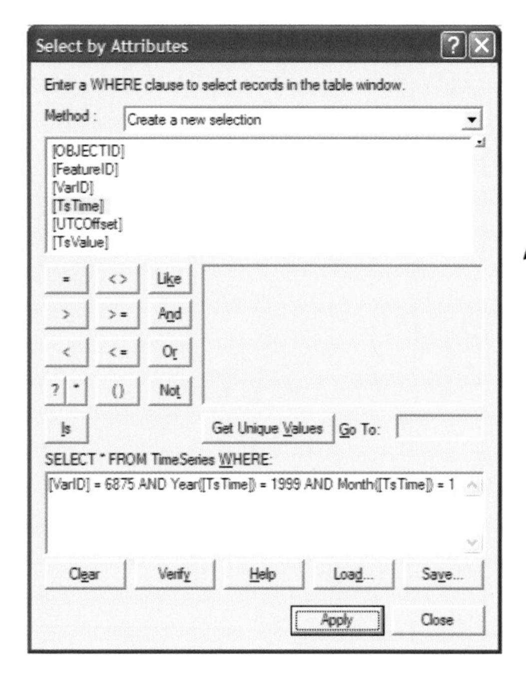

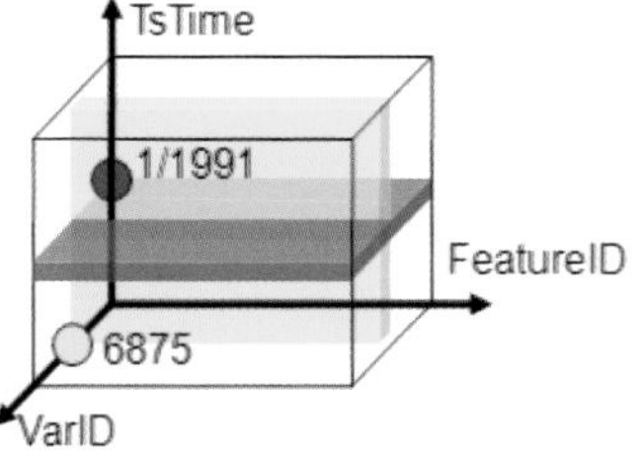

Figure 7.16 Example query for extracting a specific variable (in this case groundwater levels, VarID=6875) for a specified time (in this case, January 1991).

the same feature and time. The term attribute series was coined by combining the ideas that features have *attributes* (the attribute values that describe them in a Feature Attribute Table), and "time has *series* (the values that are indexed against each point in time in a series). Hence, if we think of feature attributes that no longer have fixed values but instead are expressed as time series, we create attribute series rather than simply attributes for each feature in a feature class. Each row in an attribute series table is indexed by a particular FeatureID and a particular TsTime. Because a single row in this table can store values for several different time-series variables, this type of table may also be known as a multivariable time-series table. For example, the Texas Water Development Board groundwater database stores results of water-quality analyses sampled at wells across Texas. For each sample analyzed, a set of about twenty water-quality parameters is recorded (e.g., TDS, pH, sulfate, calcium). All these parameters are related to the same sample, so they all can be indexed by the same feature and time. Thus, to store these data for a single feature and date/time within the single-variable TimeSeries table, we would have to populate twenty rows, one for each variable measured. A more efficient way to archive these data is to add attributes for storing multiple values for each row in the table. The result is a table structure similar to the TimeSeries

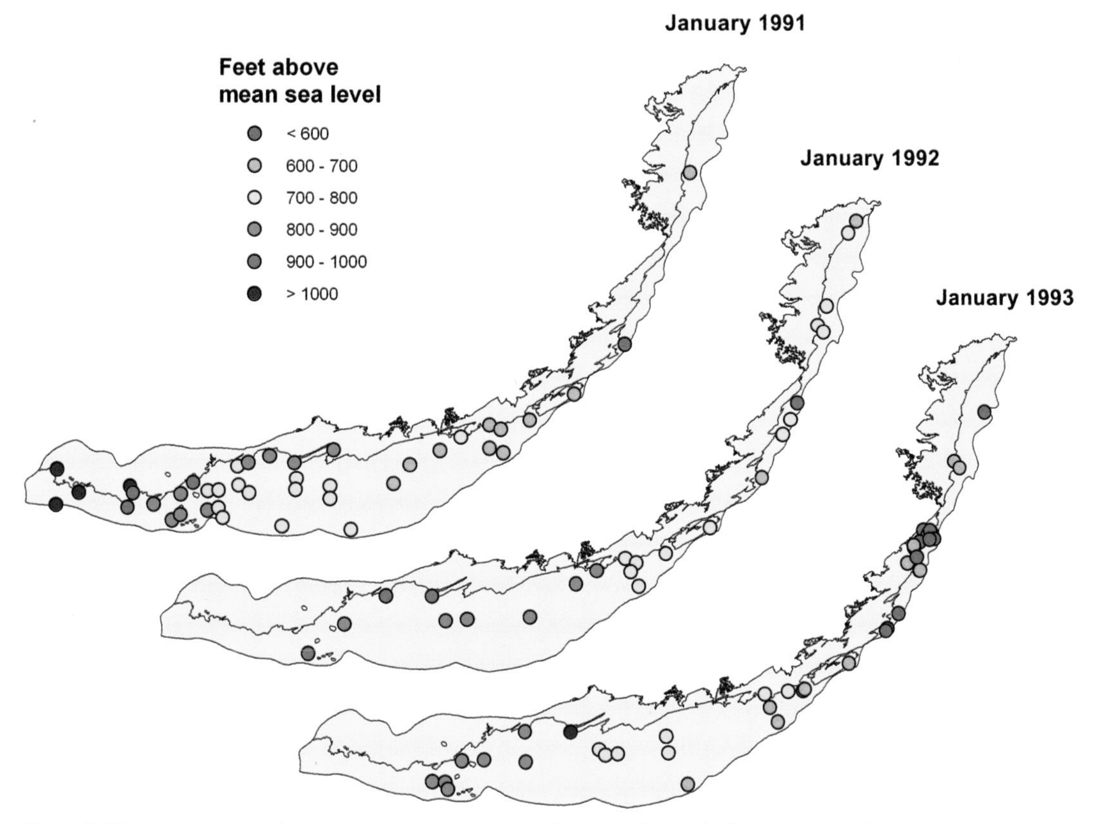

Figure 7.17 A set of animated maps showing water levels in the Edwards Aquifer for the month of January in 1991, 1992, and 1993.

table: data are indexed by space (FeatureID) and by time (TsTime), but instead of one variable, we store multiple variables. The resulting table (figure 7.18) is an attribute-series table (in this case the table is named WaterQuality to reflect that the series represent water-quality data). For a given value field, the field name indicates the variable whose TsValues are stored in that field. We call this field name the variable key (VarKey). Each of these variables is described in the VariableDefinition table as before, with the linkage between VariableDefinition and the attribute series established via the VarKey. Thus, the VarKey serves the same role as the VarID from the single-attribute time-series table, except that VarKey is better suited to be used as a field name as required by the attribute-series table.

We analyze attribute series the same way we analyze single-variable time series. First we query for data related with a feature of interest over a specified time frame, then we can select one or more of the variables stored in the attribute-series table for inclusion in the time-series plot. For example, figure 7.19 shows a plot of water-quality parameters related with well 2833 located near Comal Springs in New Braunfels, Texas. The plot was created by querying an attribute-series table of water quality for FeatureID = 2833.

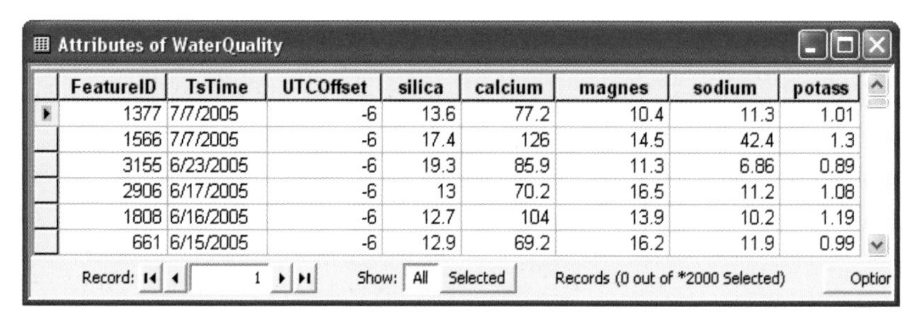

Attributes of WaterQuality

FeatureID	TsTime	UTCOffset	silica	calcium	magnes	sodium	potass
1377	7/7/2005	-6	13.6	77.2	10.4	11.3	1.01
1566	7/7/2005	-6	17.4	126	14.5	42.4	1.3
3155	6/23/2005	-6	19.3	85.9	11.3	6.86	0.89
2906	6/17/2005	-6	13	70.2	16.5	11.2	1.08
1808	6/16/2005	-6	12.7	104	13.9	10.2	1.19
661	6/15/2005	-6	12.9	69.2	16.2	11.9	0.99

Record: 1 Show: All Selected Records (0 out of *2000 Selected) Optior

Figure 7.18 Example of an attribute-series table storing water-quality data.

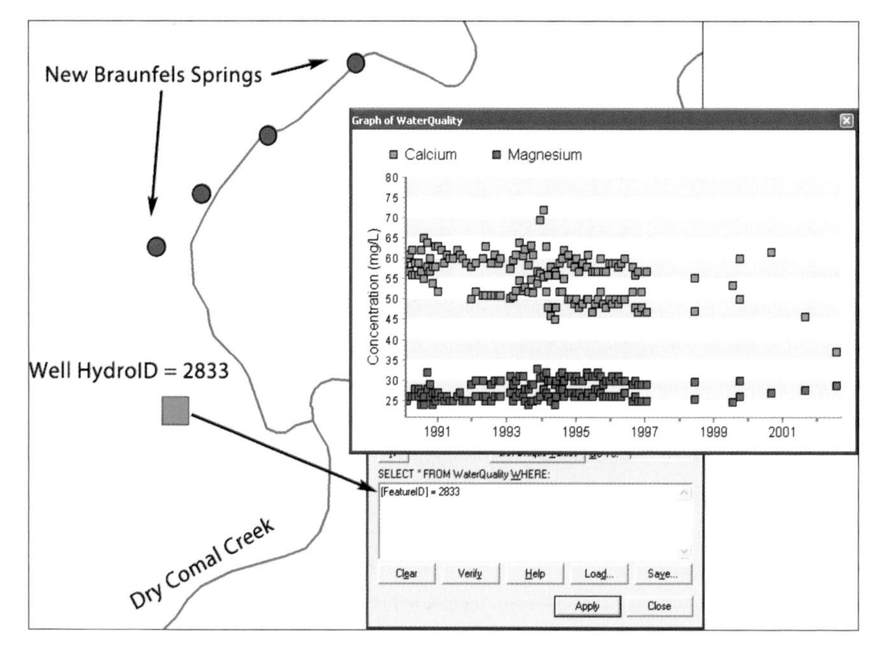

Figure 7.19 Plot of water quality at a well near Comal Springs in New Braunfels, Texas. The plot was created by querying an attribute-series table containing water-quality measurements. By querying for a specific feature one can plot multiple water-quality time series.

Raster series

The above examples have demonstrated how time-series datasets are stored within the Arc Hydro geodatabase. The task of mapping and animating time-series data stored in single- and multi-variable time-series tables is common to many groundwater studies. For analysis purposes it is also common to create continuous datasets describing the spatial distribution of a variable at a given time. These datasets are usually represented as surfaces interpolated from monitoring data and stored as raster datasets.

RasterSeries is a raster catalog for storing collections of raster datasets indexed by time (figure 7.20). Each raster is a snapshot of the environment at some instant in time, and grouping a series of rasters describes how the environment changes over time. Raster series are useful for describing the dynamics of spatially continuous phenomena, like the variations in groundwater levels or the distribution of rainfall over time.

RasterSeries

Raster catalog for storing collections of raster datasets indexed by time

Time 1
Time 2
Time 3

Field name	Description
Name	Generic attribute (created automatically when you create a raster catalog) used to store the name of a raster dataset.
VarID	Numerical identifier for the variable within the geodatabase. Matches the HydroID of the associated record in the VariableDefinition table.
TsTime	Time stamp specifying the date and time associated with the raster.
UTCOffset	Number of hours the time coordinate system used to define TsTime is displaced from Coordinated Universal Time.

Figure 7.20 RasterSeries raster catalog for storing collections of time-indexed rasters.

Rasters in the RasterSeries catalog are indexed by time (TsTime and UTCOffset) and by the time-series variable (VarID). Thus, each record in the RasterSeries catalog describes the continuous distribution of a single variable at a given time. For example, the water-level data shown for the month of January in 1991, 1992, and 1993 can be interpolated into three water-level surfaces represented as raster datasets (figure 7.21). Similar to the water-level point features, rasters in the RasterSeries catalog can be animated over time to show the spatial-temporal change of a certain variable.

Feature series

The previous examples have described data structures where time-varying data is related with stationary features (e.g., wells and gages). Time-series data can also describe dynamic

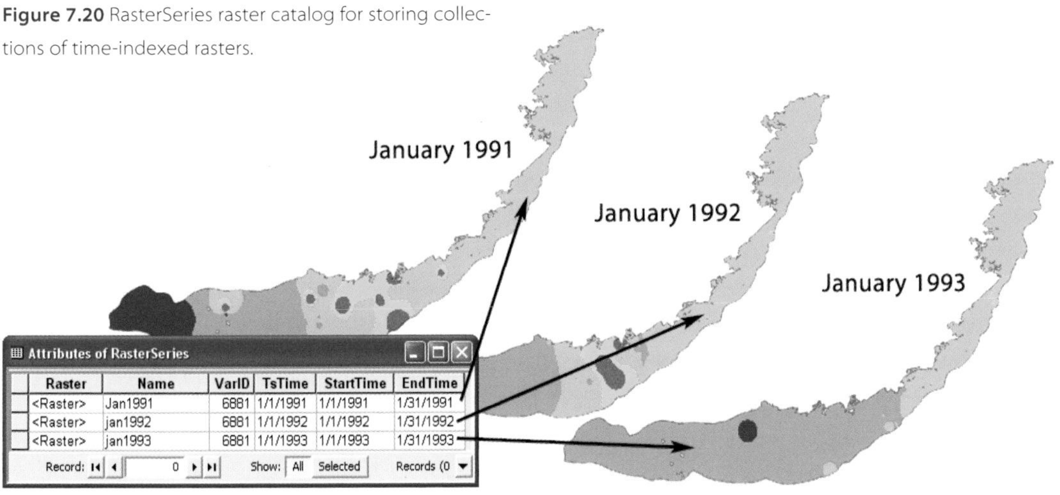

Raster	Name	VarID	TsTime	StartTime	EndTime
<Raster>	Jan1991	6881	1/1/1991	1/1/1991	1/31/1991
<Raster>	jan1992	6881	1/1/1992	1/1/1992	1/31/1992
<Raster>	jan1993	6881	1/1/1993	1/1/1993	1/31/1993

Record: 0 Show: All Selected Records (0

Figure 7.21 Rasters representing water-level surfaces within the Edwards Aquifer stored in a RasterSeries raster catalog.

features where the feature itself is moving over time or changing its geometry. A common example is the use of time-varying features to create particle tracks that describe the movement of water and the time of travel of constituents within an aquifer. To describe such data, Arc Hydro introduces the concept of feature series.

FeatureSeries are collections of features indexed by time representing a series of geometries varying in location or shape. Each feature in a feature series exists for only a period of time. Features in a feature series can be grouped to form a "track" representing the change (in geometry, location, or both) of a particular feature over time. Figure 7.22 shows the attributes of a feature-series dataset.

Take for example the particle tracks shown in figure 7.23. Each feature in the dataset is indexed by VarID and TsTime, and in addition to these standard indexes the particles are also given a GroupID that creates a "particle track" from a set of standalone particles (in this example, four

FeatureSeries

Collections of features indexed by time representing a series of geometries varying in location or shape

Time 1
Time 2
Time 3

Field name	Description
HydroID	Unique feature identifier in the geodatabase used for creating relationships between classes of the data model.
VarID	Numerical identifier for the variable within the geodatabase. Matches the HydroID of the associated record in the VariableDefinition table.
TsTime	Time stamp specifying the date and time associated with the feature.
UTCOffset	Number of hours the time coordinate system used to define TsTime is displaced from Coordinated Universal Time.
GroupID	Index for grouping a set of features into a "track" showing the change of a feature's shape, location, or both.

Figure 7.22 Feature series are collections of features that represent the change of a feature's shape, location, or both.

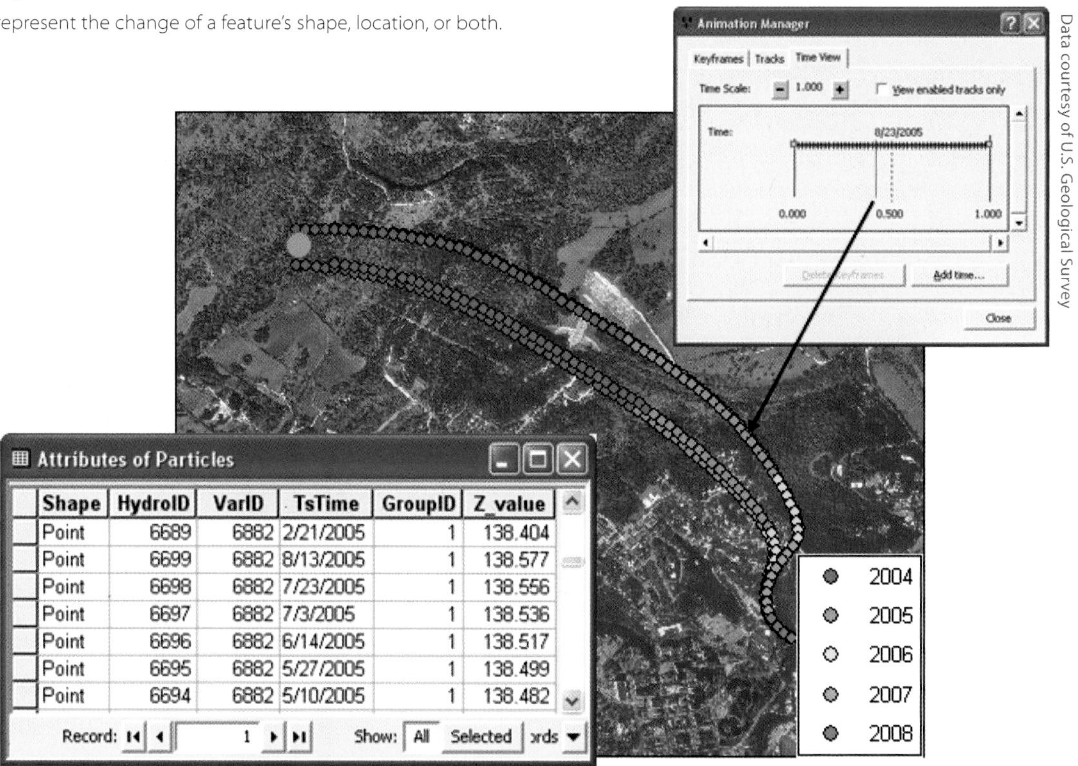

Shape	HydroID	VarID	TsTime	GroupID	Z_value
Point	6689	6882	2/21/2005	1	138.404
Point	6699	6882	8/13/2005	1	138.577
Point	6698	6882	7/23/2005	1	138.556
Point	6697	6882	7/3/2005	1	138.536
Point	6696	6882	6/14/2005	1	138.517
Point	6695	6882	5/27/2005	1	138.499
Point	6694	6882	5/10/2005	1	138.482

Figure 7.23 Animation of feature series representing the movement of particles through space and time.

tracks are indexed from 0 to 3). In addition to these indices, features in a feature series can be attributed with time-varying parameters. In this case, particles are attributed with the depth at which they are observed below the ground surface (Z_value). The combination of the Shape, TsTime, and GroupID supports tracking the movement of the time-varying particle features through space and time. We can create views of the feature series at different times or simply animate the track.

Dataset catalog

Recall that the SeriesCatalog table indexes individual time series in the TimeSeries table. Similarly, the DatasetCatalog table indexes time-series datasets within a geodatabase, except that entries in the DatasetCatalog relate to time-series datasets as a whole rather than an individual feature. Whereas the SeriesCatalog applies to the TimeSeries table, the DatasetCatalog is typically associated with RasterSeries or FeatureSeries in the geodatabase. For example, one entry in the DatasetCatalog might indicate that groundwater levels are represented in the geodatabase as a raster series with X number of time steps. Figure 7.24 shows the structure of the DatasetCatalog table.

Figure 7.25 shows an example of a DatasetCatalog table summarizing two time-series datasets. The selected row in the table shows a time series of VarID 6882, which in the VariableDefinition table is defined as subsurface particle tracks. DsType indicates that this is a feature series type dataset, and the source of the data is the Particles feature class. The start and end dates show the time range of the dataset, and the StepCount shows the number of date/time stamps (unique TsTime values) in the dataset. The DatasetCatalog and SeriesCatalog provide summary tables that enable us to quickly understand what types of time series are available within the geodatabase, the time period they cover, and in what type of datasets they are represented.

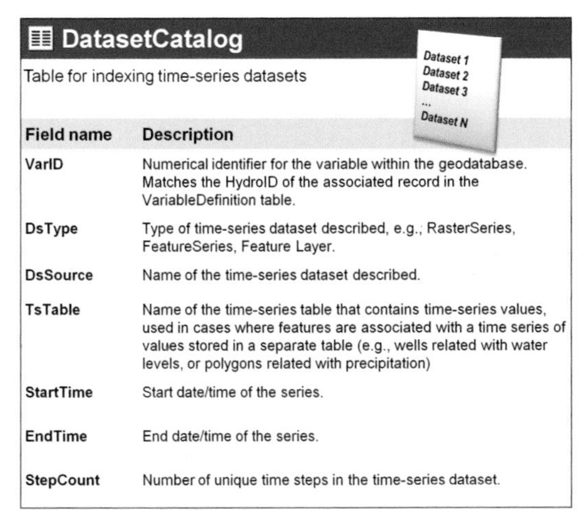
DatasetCatalog
Table for indexing time-series datasets

Field name	Description
VarID	Numerical identifier for the variable within the geodatabase. Matches the HydroID of the associated record in the VariableDefinition table.
DsType	Type of time-series dataset described, e.g.; RasterSeries, FeatureSeries, Feature Layer.
DsSource	Name of the time-series dataset described.
TsTable	Name of the time-series table that contains time-series values, used in cases where features are associated with a time series of values stored in a separate table (e.g., wells related with water levels, or polygons related with precipitation)
StartTime	Start date/time of the series.
EndTime	End date/time of the series.
StepCount	Number of unique time steps in the time-series dataset.

Figure 7.24 DatasetCatalog table for indexing time-series datasets (e.g., FeatureSeries and RasterSeries).

Time enablement of ArcGIS layers

Beginning with ArcGIS version 10, feature classes and tables can be "time-enabled" by specifying the field in the attribute table athat contains the time stamp and specifying some properties of this time

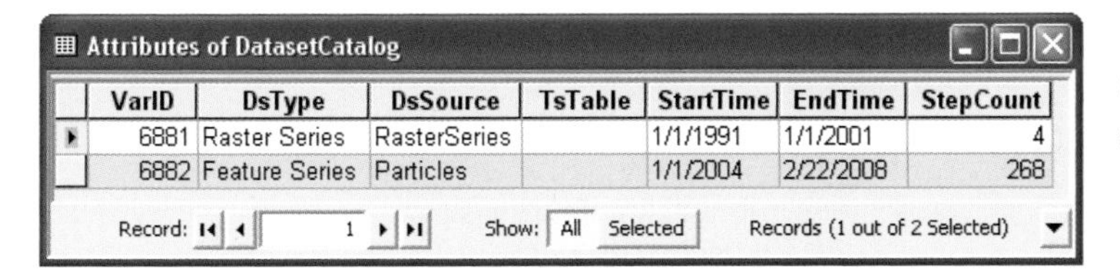
Attributes of DatasetCatalog

VarID	DsType	DsSource	TsTable	StartTime	EndTime	StepCount
6881	Raster Series	RasterSeries		1/1/1991	1/1/2001	4
6882	Feature Series	Particles		1/1/2004	2/22/2008	268

Figure 7.25 DatasetCatalog table summarizing time-series datasets.

index, for example, its time zone and time interval. Functions are provided in ArcToolbox to convert data from one time zone to another. Graphing data indexed against different time zones is automatically handled in ArcGIS so that the information is viewed in a single-time context. This is particularly important when merging temporal information from weather and climate data sources with information from regular water data sources. Weather and climate data are typically indexed using the Universal Time Coordinate system, applicable at Greenwich, England, for data from anywhere on earth. This is done so that worldwide weather maps can be drawn for a unique point in time. Because groundwater and water resources data in general are usually viewed locally rather than globally, local time coordinates are typically used to record the time of measurement.

Time-enablement of ArcGIS layers means that a record in a feature attribute table can have a time stamp associated with it, and if other records in this table describe the same feature at different times, or other features at the same time, then the beginnings of a space-time analysis system exists. Time-enablement of ArcGIS data structures enhances and strengthens the temporal component of Arc Hydro.

Framework and extended temporal components

When integrating time-series data with a GIS, a common approach is to first collect single-variable time series of interest, and then use GIS tools to derive interpreted products such as raster series or feature series. Therefore, the Arc Hydro temporal component is available in two levels: *framework* and *extended*. The framework version includes the TimeSeries, VariableDefinition, and SeriesCatalog table. These tables store, describe, and index single-variable time series, respectively. The extended version includes all the tables from the framework version, plus RasterSeries, FeatureSeries, AttributeSeries, and DatasetCatalog. If you do not need these additional datasets, then the framework version provides a lightweight geodatabase design for your work. Otherwise, the extended version gives you the full capabilities of the Arc Hydro temporal component.

Steps for implementing the temporal component

This section outlines the main steps involved in archiving and analyzing time-series data using the temporal component datasets. The first steps involved in implementing this component are to refine the geodatabase design to meet your specific project needs (see chapters 1 and 9 for a detailed description of these steps). Then you can import data into the time-series component tables and start creating different space-time datasets. In the following example we use a common workflow, creating water-level maps from water-level measurements taken at wells, for demonstrating the process of working with time-series data. The first step is to import time-series data into the tabular structures and then assign key attributes to relate time-series records with spatial features. You can then apply tools to create new space-time datasets such as maps of time series averaged over specific time periods and raster datasets created by interpolating time series measured at point locations. Finally, you can use the results of this process to create products such as maps, scenes, graphs, and animations. The following checklist provides a summary of the main steps for implementing the temporal component.

Checklist

1. Create the classes of the temporal component (manually using ArcCatalog or by importing from an XML schema)
2. Add project specific classes, attributes, relationships, and domains as necessary
3. Document datasets and changes made to the data model
4. Import time-series data into the appropriate tables (TimeSeries, AttributeSeries)
5. Establish relationships between time series and spatial features by populating key attributes
6. Apply tools to create new time-series datasets (e.g., average values over a specified time period, interpolate raster datasets)
7. Populate the summary catalog tables (SeriesCatalog, DatasetCatalog)
8. Visualize time-series data and create products (maps, scenes, plots, animations)

Attributes of TimeSeries

FeatureID	VarID	TsTime	UTCOffset	TsValue
2200	6875	9/8/1881	-6	564.81
2209	6875	7/7/1882	-6	598.93
2677	6875	11/30/1918	-6	677
410	6875	3/12/1922	-6	714.7
2648	6875	11/30/1922	-6	639
1395	6875	9/20/1929	-6	836.8
1539	6875	10/2/1929	-6	862.8
1486	6875	10/21/1929	-6	860.95
1549	6875	10/24/1929	-6	856.95

Record: 1 Show: All Selected :ords

Figure 7.26 Example of water-level time series imported into the TimeSeries table.

We start with importing our time-series data from its original form (text files, spreadsheets, database tables) into the table structures of the temporal component. In the following example (figure 7.26), water-level data from the Texas Water Development Board's groundwater database were imported into the TimeSeries table. Key attributes were assigned after importing the data into the table. For example, water-level records were attributed with FeatureID values to associate the records with Well features. The original table in the groundwater database indexes wells by a "state well ID" (in the Arc Hydro geodatabase, this is stored in the HydroCode attribute). While this index is valid, the HydroID is used in the Arc Hydro geodatabase to assure that each feature is uniquely identified within the geodatabase. Thus, each of the water-level records was assigned a FeatureID corresponding to the well where the water levels were recorded. Also, each of the water-level records was attributed with a VarID (in this case VarID = 6875 for water levels), and UTCOffset of -6 to represent the local time zone in Central Texas. The water-level records and the time values were imported into the TsTime and TsValue fields.

The next step is to apply tools to create new temporal datasets from the original time series. It is common to calculate summary statistics of a time-varying variable to represent the variable over a given time period. For example, we might want to create a water-level map for the winter of 2001, such that each well will have a single value representing the winter period (January through March). There could be multiple measurements taken at a well during this period, thus we need to calculate a summary statistic that represents the water level at a well for the required time period. For example, we can select the representative value as the average of all water levels measured between January 1 and March 31, 2001 (the Arc Hydro Groundwater tools include a geoprocessing tool, Make Time Series Statistics, to support this task). Figure 7.27 shows the result of the averaging process, a new feature class where each feature includes an attribute giving the summary statistic calculated from the time-series records.

The point features with the calculated statistics can be the basis for interpolating water-level rasters. For example, attribute values of the point features shown in figure 7.27 were used as inputs to the IDW interpolation tool (available with the Spatial Analyst extension). The result is a raster dataset representing the mean water level within the aquifer during the winter of 2001. The raster dataset can then be loaded into the RasterSeries raster catalog and indexed with a VarID and time stamped with a TsTime (figure 7.28). The Arc Hydro Groundwater tools also include a tool, Add to Raster Series, to support automatic loading and attribution of rasters into the RasterSeries raster catalog.

This workflow of going from water-level measurements in the time-series table to creating raster datasets stored and attributed in the RasterSeries raster catalog can be automated with a

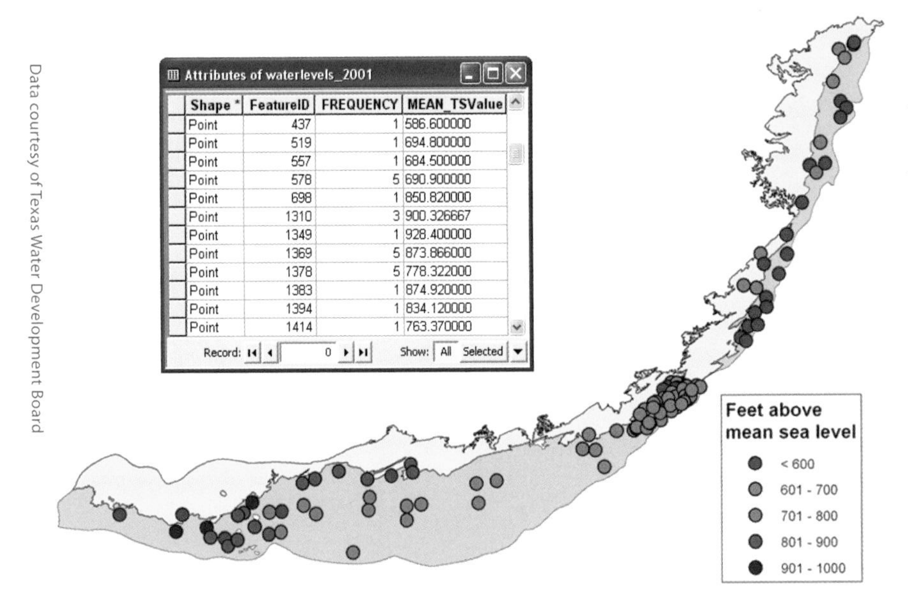

Figure 7.27 Averaged water levels for wells in the Edwards Aquifer between January 1 and March 31, 2001. The Make Time Series Statistics tool was used to create the new feature class representing the summary statistics over the selected time period.

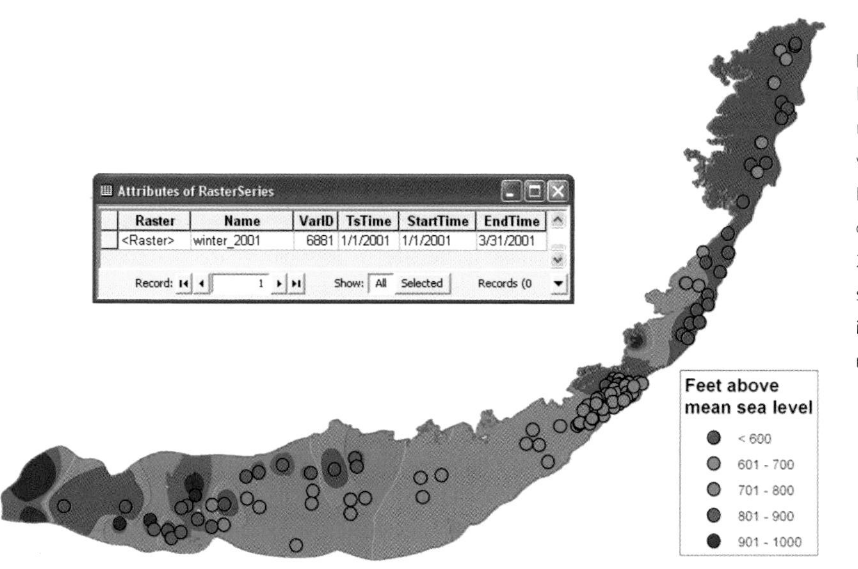

Figure 7.28 Interpolated raster dataset representing average water levels in the Edwards Aquifer during the winter of 2001. The raster is stored and attributed in the RasterSeries raster catalog.

single model as shown in figure 7.29. The model starts with the Make Time Series Statistics tool that takes time series stored in the TimeSeries table and creates a new set of point features with summary statistics values for a defined time period. The time-series statistics values are then interpolated to create a raster representing the water level as a continuous surface, and the raster is indexed and stored in the RasterSeries raster catalog using the Add to Raster Series tool.

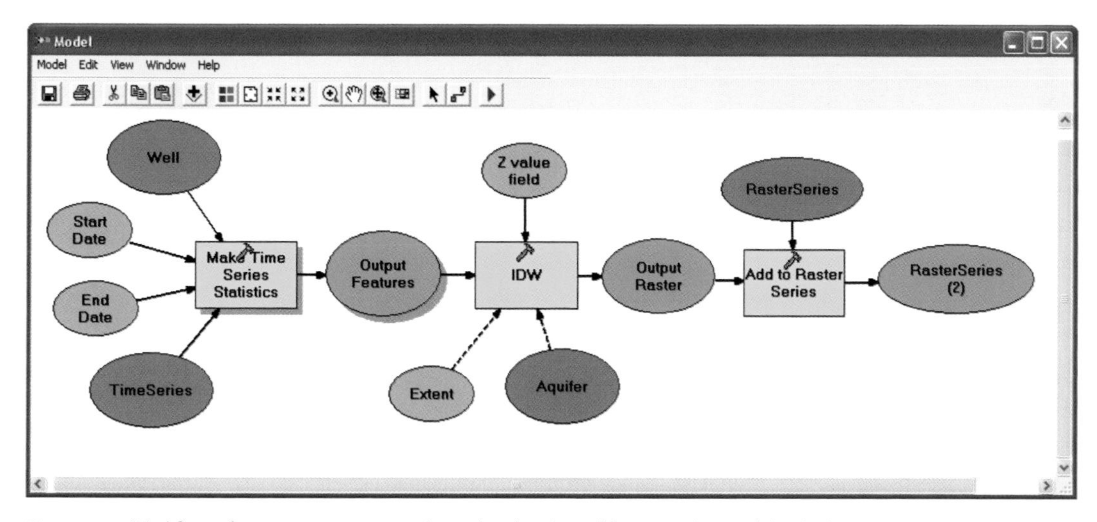

Figure 7.29 Workflow of mapping time-series data related with Well features. The model includes creating time-series statistics, interpolating the calculated values to a raster, and storing and attributing the raster in the RasterSeries raster catalog.

8 chapter eight

Groundwater simulation models

NORMAN L. JONES AND GIL STRASSBERG

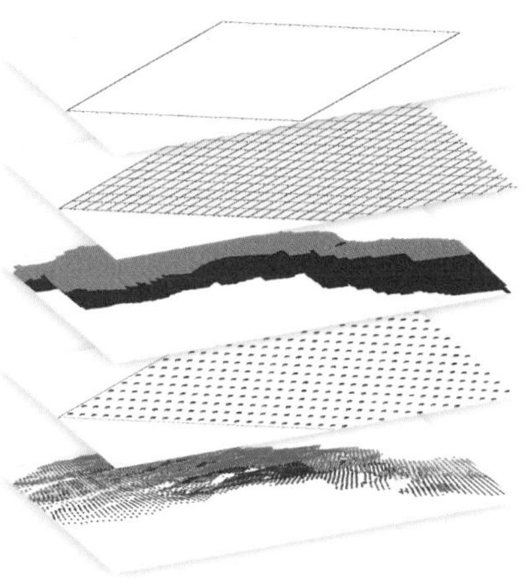

Boundary
Defines the 2D extent and orientation of a simulation model.
Representation: Polygon features.

Cell2D
Represents cells or elements of a 2D simulation model or a single layer of a 3D model.
Representation: Polygon features.

Cell3D
3D features representing cells or elements of a simulation model.
Representation: Multipatch features.

Node2D
Represents the computational nodes of a 2D simulation model or a singe layer in a 3D model.
Representation: Point features.

Node3D
3D points representing the computational nodes of a simulation model.
Representation: PointZ features.

ONE OF THE PRIMARY USES OF THE ARC HYDRO GROUNDWATER DATA MODEL IS to support integration of groundwater simulation models with GIS. Simulation models are used to build simplified mathematical representations of complex aquifer systems, to interpret the flow of water and transport of contaminants within an aquifer, and to predict how these will change under future stresses. Simulation models are used to assist water-resource managers in making decisions, balance the competing needs of groundwater stakeholders, predict the fate and migration of groundwater pollutants, and design strategies for remediating contaminated aquifers. Simulation models typically involve some type of numerical grid where the model inputs are discretized into values associated with grid cells. We can use standard GIS functionality to automate the process of preparing model inputs by discretizing spatial data layers (e.g., land cover, soils, elevation, and aquifer thickness) over the model grid. GIS also is ideal for organizing, archiving, and presenting results of simulation models. The results can be mapped in context with other spatial datasets to convey model results to decision makers and the public. Once stored within a geodatabase, model inputs and outputs can be served over the Internet in a simple format usable by any GIS user.

In this chapter we review the more common types of simulation models and describe how groundwater models can be represented in the simulation component of the groundwater data model. We present the concept of interface data models, which are extensions developed to store complete instances of simulation models. We also present an example of an interface data model, called the MODFLOW (modular finite-difference flow model) data model, for storing MODFLOW models.

Types of simulation models

Most groundwater simulation models fall into one of three categories: analytic element, finite element, and finite difference. For the analytic element method, principal features of aquifer systems such as rivers, flow barriers, zones of hydraulic conductivity, and wells are represented using points, lines, and polygons (figure 8.1). A series of equations is then derived using principles of superposition, and a set of "strength coefficients" associated with the equations is computed. Once these coefficients are determined, the principal equations can be solved to simulate the water table elevation and flow quantities at any x,y location. The analytic element method has advantages over the finite element and finite difference methods in that there is no need to discretize the model domain. On the other hand, it requires more simplifying assumptions than the other two methods and cannot be applied to certain situations involving complex geometry or heterogeneity.

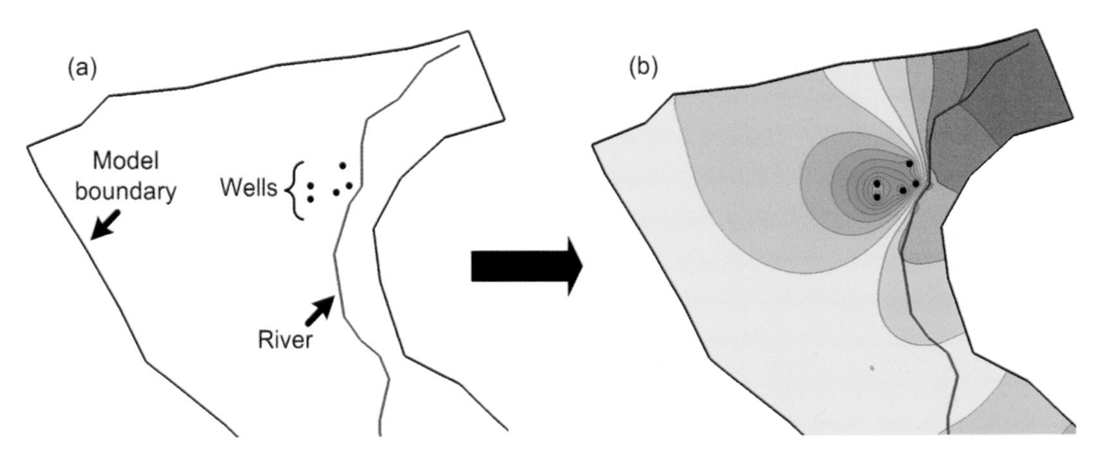

Figure 8.1 Analytic element model of an aquifer flow system: (a) model inputs such as rivers, wells, and the model boundary are defined via GIS features; and (b) head contours are generated by solving the analytic element equations.

With the finite element method, the model domain is subdivided or discretized into small units called "elements." For 2D simulations, the elements consist of triangle and quadrilateral (rectangular) shapes. For 3D simulations, the elements take the form of tetrahedra, prisms, hexahedra ("bricks"), or pyramids (figure 8.2). Properties such as permeability are assigned to the elements, and boundary conditions corresponding to aquifer stresses including wells, rivers, and drains are established. An equation is then established for each element that is related to the physics of the system being modeled. The element equations include the effect of the material properties, geometry, and boundary conditions. The element equations are combined into a single global matrix, and the matrix is solved. The solution to the matrix consists of water table elevations and flow rates over the entire model domain. The finite element method can be used to simulate extremely complex geometries, but it is the most complex and computationally intensive of the three primary simulation methods.

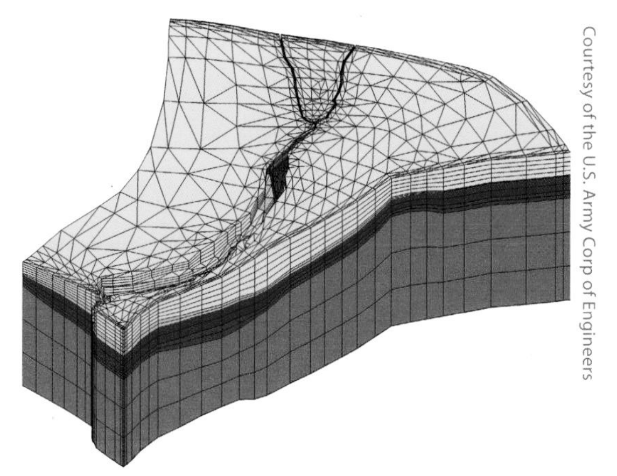

Courtesy of the U.S. Army Corp of Engineers

Figure 8.2 An example of a 3D finite element groundwater model.

The most common type of groundwater simulation model is the finite difference method. With the finite difference method the model domain is represented using an orthogonal grid organized into rows and columns, and each rectangular unit of the grid is called a cell. For 3D simulations, layers are added to the model, and the cells are in the shape of regular hexahedra. Material properties and aquifer stresses are assigned to each of the cells in the grid. As in the case of the finite element method, a global set of equations is established and solved for the distribution of water levels and flow rates. The finite difference method is relatively simple and efficient and is the basis for the most widely used groundwater simulation model, MODFLOW, developed by the USGS (Harbaugh et al. 2000).

Geodatabase representation of simulation models

The simulation component includes a set of feature classes that represent common modeling objects of finite element and finite difference models. The component includes five feature classes: Boundary, Cell2D, Cell3D, Node2D, and Node3D (figure 8.3).

Boundary is a polygon feature class that represents the 2D extent of a simulation model. It is not an essential part of the numerical model representation, but it is useful for mapping the spatial domain of the model. For a finite-difference grid the Boundary feature would be in the shape of a rectangle, and for a finite element mesh it would be an irregular shape representing the outer perimeter of the model. The Boundary feature class can include multiple polygon features that represent multiple models. This allows the development of a single layer to represent the extent of multiple simulation models. The attributes of Boundary features are shown in figure 8.4.

The geographic orientation of a model is defined by the model's origin and the angle of rotation. In the following example (figure 8.5), the angle of rotation φ is defined as the angle between the north and the j axis (the axis defining columns in a MODFLOW model). The angle can be defined in various ways, thus how the angle is calculated

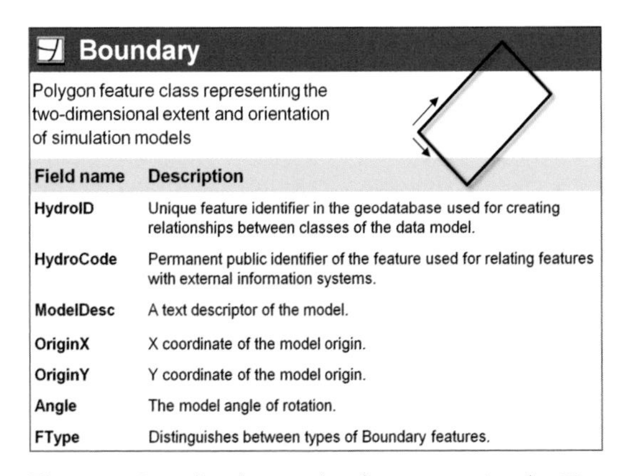

Boundary

Polygon feature class representing the two-dimensional extent and orientation of simulation models

Field name	Description
HydroID	Unique feature identifier in the geodatabase used for creating relationships between classes of the data model.
HydroCode	Permanent public identifier of the feature used for relating features with external information systems.
ModelDesc	A text descriptor of the model.
OriginX	X coordinate of the model origin.
OriginY	Y coordinate of the model origin.
Angle	The model angle of rotation.
FType	Distinguishes between types of Boundary features.

Figure 8.4 Boundary feature class for representing the 2D extent of simulation models.

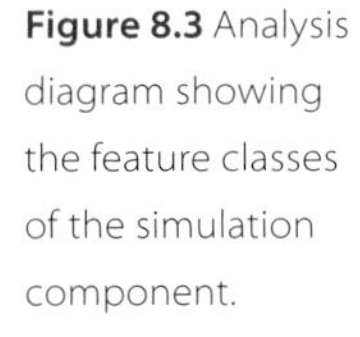

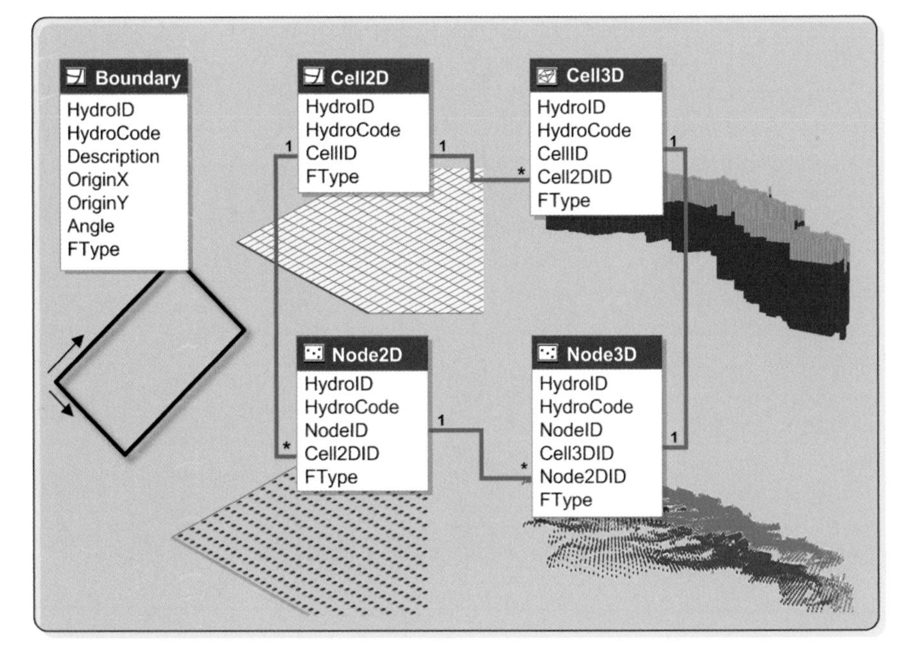

Figure 8.3 Analysis diagram showing the feature classes of the simulation component.

should be documented in the metadata of the feature class.

Cell2D is a polygon feature class that represents cells or elements associated with a 2D simulation model or a single layer of a 3D model. To represent a 2D finite element mesh, Cell2D features would have a triangular or rectangular geometry representing the elements of the model, and in a finite difference model Cell2D features would be rectangular. Node2D is a point feature class that represents the computational nodes in a simulation model (i.e., the locations at which the solution is computed). Node2D features are used in combination with Cell2D features to represent the model's mesh/grid. Figure 8.6 shows different options for using Cell2D and Node2D features. In a finite element model Node2D features would be located at the corners of the elements. For finite difference models there are two cases: Node2D features would correspond to cell corners for a mesh-centered grid and to cell centers for the more common cell-centered grid.

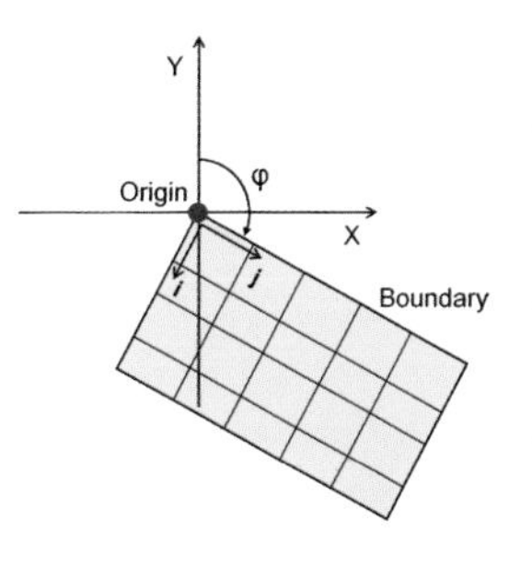

Figure 8.5 Boundary features represent the spatial extent and orientation of a simulation model. The spatial orientation of the model is defined by the origin and angle of rotation.

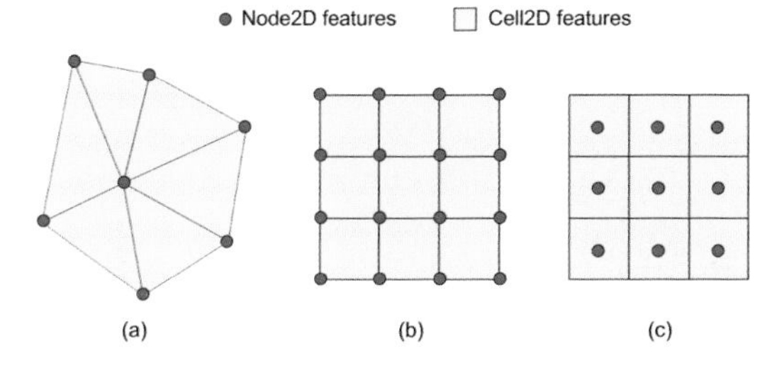

Figure 8.6 Node2D and Cell2D features for common simulation grid types: (a) finite element, (b) mesh-centered finite difference, and (c) cell-centered finite difference.

Cell2D and Node2D features are used to represent 2D models, but can also be used in combination with 3D models. For 3D models, the 3D features represent the cells and nodes used by the simulation model, and the Cell2D and Node2D features are used to map either summary information (e.g., maximum concentration) from the 3D features or data associated with a selected layer of the 3D model grid. In some cases such as linked surface water-groundwater models, both a 2D grid and a 3D grid are used in the simulation. Attributes of Cell2D and Node2D features are shown in figure 8.7 and figure 8.8, respectively.

Simulation model properties, input data, and outputs could be stored directly on the Cell2D

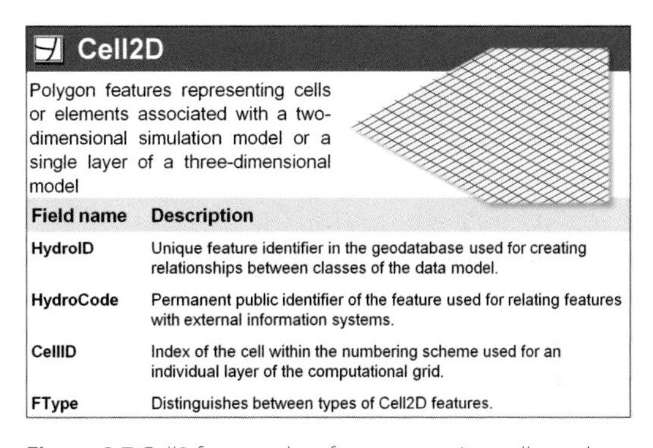

Cell2D

Polygon features representing cells or elements associated with a two-dimensional simulation model or a single layer of a three-dimensional model

Field name	Description
HydroID	Unique feature identifier in the geodatabase used for creating relationships between classes of the data model.
HydroCode	Permanent public identifier of the feature used for relating features with external information systems.
CellID	Index of the cell within the numbering scheme used for an individual layer of the computational grid.
FType	Distinguishes between types of Cell2D features.

Figure 8.7 Cell2 feature class for representing cells or elements in a 2D simulation model or a single layer of a 3D model.

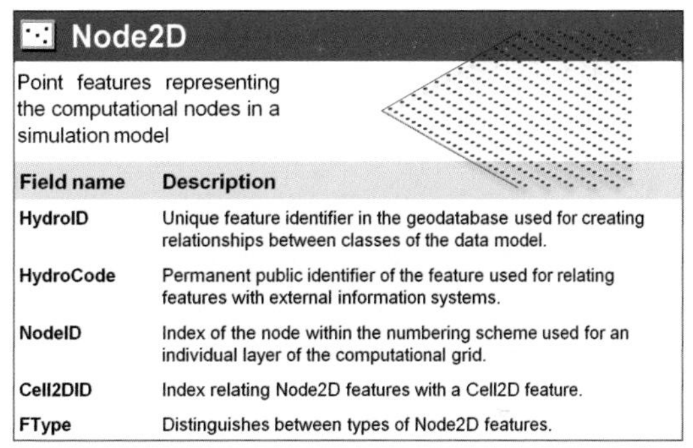

Node2D

Point features representing the computational nodes in a simulation model

Field name	Description
HydroID	Unique feature identifier in the geodatabase used for creating relationships between classes of the data model.
HydroCode	Permanent public identifier of the feature used for relating features with external information systems.
NodeID	Index of the node within the numbering scheme used for an individual layer of the computational grid.
Cell2DID	Index relating Node2D features with a Cell2D feature.
FType	Distinguishes between types of Node2D features.

Figure 8.8 Node2D feature class for representing the computational nodes in a 2D simulation model or a single layer in a 3D model.

and Node2D features, or they could be stored in separate tables and related to the features by association with the HydroID or the CellID. The relationship between Cell2D and Node2D is established through the Cell2DID-CellID relationship (figure 8.9). Each Node2D feature is indexed with a Cell2DID that is equal to the CellID of a Cell2D feature. For cell-centered finite difference grids, the relationship between cells and nodes is a one-to-one relationship since each cell is associated with a single node. For mesh-centered finite difference and finite element models, a many-to-many approach is required as nodes can be related with multiple cells, and cells are associated with multiple nodes. In this case an intermediate table will be developed relating CellID and NodeID values in a many-to-many relationship (see chapter 9 for more details on many-to-many relationships).

A similar approach is used to represent a 3D model, but in this case the elements of the model grid are represented with 3D features. Cell3D is a multipatch feature class that represents 3D cells and elements; and Node3D is a z-enabled point feature class for representing the computational points of a 3D model grid. Cell3D and Node3D features are used mostly for visualization of 3D model data in ArcScene and less for data archiving. Attributes of Cell3D and Node3D features are shown in figure 8.10 and figure 8.11, respectively.

Similarly to the Cell2D-Node2D relationship, Cell3D and Node3D features are also related with Cell2D and Node2D features (figure 8.12). Each Cell3D feature is attributed with a Cell2DID index that relates to the CellID of a Cell2D feature. Node3D features are indexed with a Node2DID that relates 3D nodes with a Node2D feature.

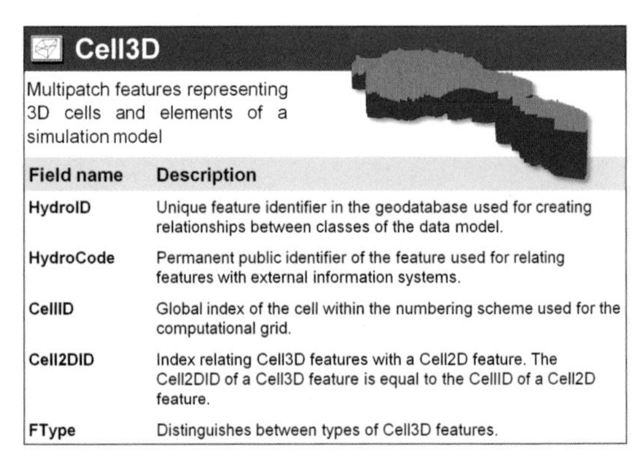
Cell3D

Multipatch features representing 3D cells and elements of a simulation model

Field name	Description
HydroID	Unique feature identifier in the geodatabase used for creating relationships between classes of the data model.
HydroCode	Permanent public identifier of the feature used for relating features with external information systems.
CellID	Global index of the cell within the numbering scheme used for the computational grid.
Cell2DID	Index relating Cell3D features with a Cell2D feature. The Cell2DID of a Cell3D feature is equal to the CellID of a Cell2D feature.
FType	Distinguishes between types of Cell3D features.

Figure 8.10 Cell3D feature class for representing 3D cells and elements of a simulation model.

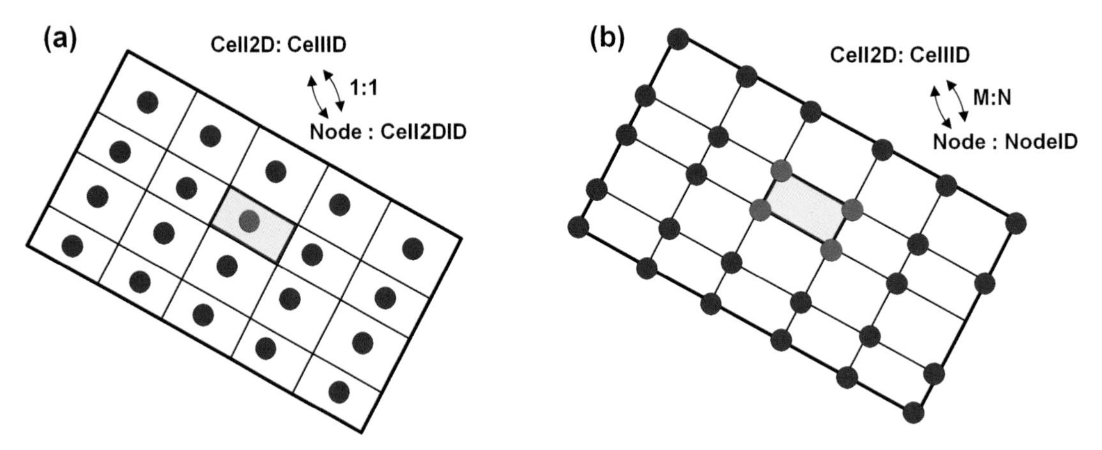

Figure 8.9 Relationship between Cell2D and Node2D features: a) Node2D features are indexed with a Cell2DID, which relates to the CellID of a Cell2D feature; and b) Cell2D and Node2D features are related through a many-to-many relationship.

As was the case with Node2D and Cell2D features, the relationship between Cell3D and Node3D features is established through the Cell3DID-CellID relationship (figure 8.13). Each Node3D feature is indexed with a Cell3DID, which is equal to the CellID of a Cell3D feature. This one-to-one relationship is used for cell-centered finite difference grids where each cell is associated with a single node. A many-to-many approach is required for mesh-centered finite difference and finite element models, as nodes can be related with multiple cells, and cells are associated with multiple nodes.

Interface data models

An Interface data model is a geodatabase design for storing the entire contents of a given simulation model. While the simulation component of the groundwater data model is general and may serve as a starting point for different types of projects and models, the objective of an interface data model is to mimic the inputs, outputs, and parameters of a numerical model, such that an entire model can be stored within the geodatabase. Storing the complete contents of a simulation model within a geodatabase provides advantages that include:

- The integration of model data within a GIS environment, which provides the capability for visualizing, editing, and preprocessing model data and parameters using standard GIS tools.
- Standard archiving format that supports queries, editing, versioning, etc.
- The capability to publish model data on the Internet.

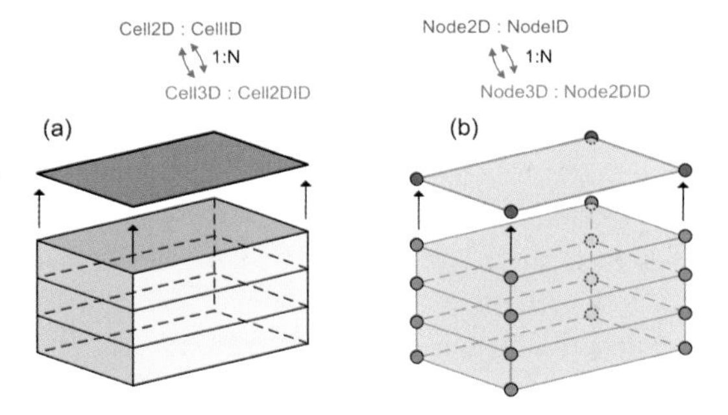

Figure 8.12 Relationship between 2D and 3D features: a) Cell3D features are indexed with a Cell2DID that relates to the CellID of a Cell2D feature; and b) Node3D features are indexed with a Node2DID that relates to the NodeID of a Node2D feature.

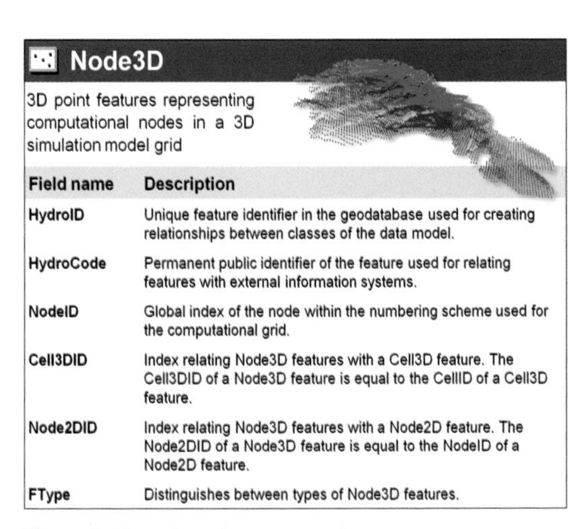

Node3D

3D point features representing computational nodes in a 3D simulation model grid

Field name	Description
HydroID	Unique feature identifier in the geodatabase used for creating relationships between classes of the data model.
HydroCode	Permanent public identifier of the feature used for relating features with external information systems.
NodeID	Global index of the node within the numbering scheme used for the computational grid.
Cell3DID	Index relating Node3D features with a Cell3D feature. The Cell3DID of a Node3D feature is equal to the CellID of a Cell3D feature.
Node2DID	Index relating Node3D features with a Node2D feature. The Node2DID of a Node3D feature is equal to the NodeID of a Node2D feature.
FType	Distinguishes between types of Node3D features.

Figure 8.11 Node3D feature class for representing computational nodes in a 3D model grid.

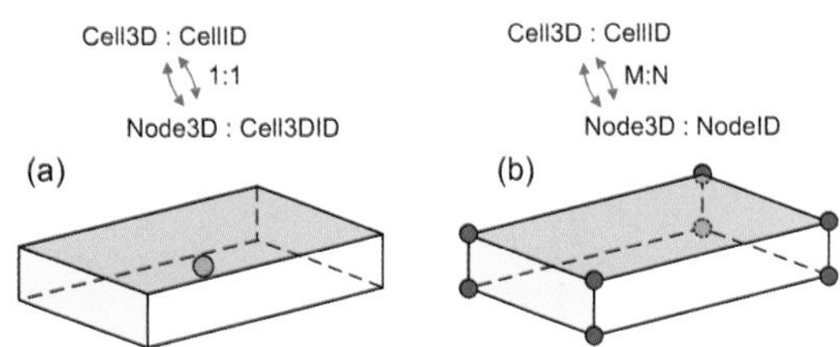

Figure 8.13 Relationship between Cell3D and Node3D features. Node3D features are indexed with a Cell3DID, which relates to the CellID of the Cell3D. The Cell3D-Node3D relationship (a) is one-to-one for cell-centered grids, and (b) the relationship is many-to-many for mesh-centered grids or finite-element meshes.

- Support for model integration by using the same underlying base datasets (e.g., the ones represented by the groundwater data model).

- The ability to build custom workflows within ArcGIS to automate tasks built on top of a simulation model (e.g., automated well permitting, capture zone analysis).

While interface data models are self-contained as they describe a complete model, different models can be based on the same underlying datasets such as aquifer boundaries, streams, and well locations. These data layers are defined in the Arc Hydro data model, and a variety of simulation models can reference the same underlying datasets through their interface data model (figure 8.14).

MODFLOW data model

The MODFLOW data model (MDM) is an interface data model developed as an extension to the Arc Hydro Groundwater data model. The MDM illustrates how the simulation feature dataset is used to support representation of groundwater simulation models. The MDM, a series of tables and relationships, supports the storage of an entire MODFLOW simulation within a geodatabase. Having the model data in the geodatabase makes it possible to use mapping and plotting capabilities of ArcGIS and to develop queries and tools, including tools for populating the MODFLOW package tables using data from the Arc Hydro components. The MDM provides an archival tool for models, facilitating scripting, permissions, check-in/check-out capabilities, and Web services.

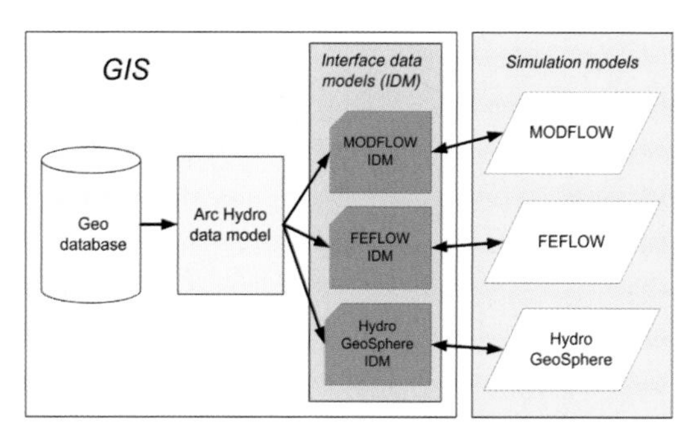

Figure 8.14 Interface data models store complete instances of simulation models within a geodatabase. A number of interface data models can be built on top of the basic datasets represented in the Arc Hydro data model.

MODFLOW tables

The bulk of MODFLOW data is stored in a series of tables. While it is impractical to present all of the tables here, a few examples are provided to demonstrate the main concepts. The tables associated with the Layer-Property Flow (LPF) package are shown in figure 8.15. Variables associated with the LPF package are stored in the LPFVars table. Layer-based variables associated with the LPF package are stored in the LPFLayers table. LPF aquifer property arrays are stored in the LPFProperties table. The table includes one record for each cell in the MODFLOW grid.

Most of the commonly used stress packages in MODFLOW follow a simple strategy. The cells associated with a given type of stress are identified, and an input entry is made for each instance in a list in the package input file. In some cases, multiple entries are made for a given cell (e.g., three

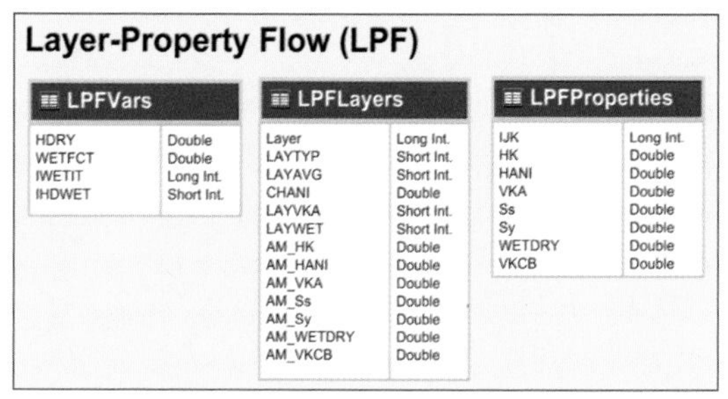

Layer-Property Flow (LPF)

LPFVars	
HDRY	Double
WETFCT	Double
IWETIT	Long Int.
IHDWET	Short Int.

LPFLayers	
Layer	Long Int.
LAYTYP	Short Int.
LAYAVG	Short Int.
CHANI	Double
LAYVKA	Short Int.
LAYWET	Short Int.
AM_HK	Double
AM_HANI	Double
AM_VKA	Double
AM_Ss	Double
AM_Sy	Double
AM_WETDRY	Double
AM_VKCB	Double

LPFProperties	
IJK	Long Int.
HK	Double
HANI	Double
VKA	Double
Ss	Double
Sy	Double
WETDRY	Double
VKCB	Double

Figure 8.15 Tables associated with the MODFLOW Layer-Property Flow (LPF) package.

wells in a single cell). The River, Well, Drain, General Head Boundary, Constant-Head Boundary, and Horizontal Flow Barrier packages follow this format. Data associated with these packages are stored in a series of tables, one per package (figure 8.16). Each record in the tables corresponds to an instance of the given stress at a particular cell for a particular time (stress period). The SPID field refers to the ID of the stress period. Data defining each stress period are stored in a stress periods table. The names of the fields that store the properties used by each package are derived from the variable names used by MODFLOW.

In a similar fashion, the MODFLOW data model includes tables for representing many of the packages used by MODFLOW. The full MDM can be viewed at (`www.archydrogw.com`).

Feature classes

The foundation of the MODFLOW data model is the Cell2D and Node2D feature classes. MODFLOW uses a cell-centered finite difference approach where the grid is organized in layers. Within each layer, the cells are ordered by rows and columns. As a result, each cell has a row (I), column (J), and layer (K) index. There is one Cell2D and one Node2D feature for each IJ-combination in the MODFLOW grid. For example, the Cell2D and Node2D features represent a single layer of the MODFLOW grid. Cell2D and Node2D features are then joined with MODFLOW data in tables to generate map layers. Node3D and Cell3D features are optional and can be used to display 3D MODFLOW data in ArcScene.

To facilitate joins and queries related to the MODFLOW data, the MDM includes a CellIndex table (figure 8.17). In the MODFLOW input files, cells are identified by a combination of three integers representing the row, column, and layer indices of the cell (I, J, K). Since joins can only be accomplished using a single field, it is necessary to generate an integer field corresponding to a unique IJK combination. Furthermore, some MODFLOW input arrays are fundamentally 2D in nature (e.g., evapotranspiration and recharge) and require an integer identifier corresponding to a unique IJ combination. The IJK field of the CellIndex table is populated from the I, J, and K values by starting at a value of one for the cell corresponding to I = 1, J = 1, and K = 1, looping through the cells in the grid row-by-row within each layer, and incrementing

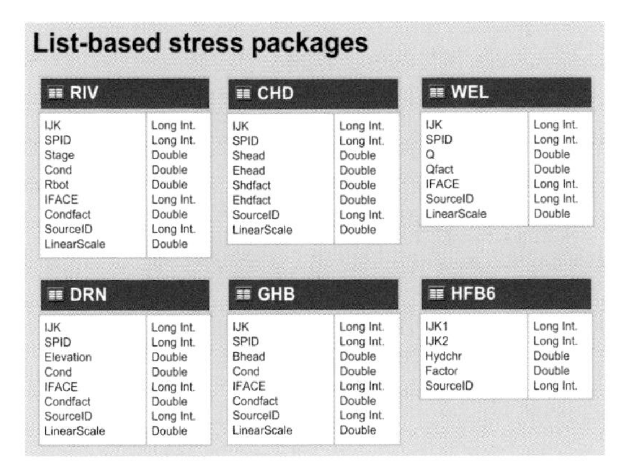
List-based stress packages

RIV

Field	Type
IJK	Long Int.
SPID	Long Int.
Stage	Double
Cond	Double
Rbot	Double
IFACE	Long Int.
Condfact	Double
SourceID	Long Int.
LinearScale	Double

CHD

Field	Type
IJK	Long Int.
SPID	Long Int.
Shead	Double
Ehead	Double
Shdfact	Double
Ehdfact	Double
SourceID	Long Int.
LinearScale	Double

WEL

Field	Type
IJK	Long Int.
SPID	Long Int.
Q	Double
Qfact	Double
IFACE	Long Int.
SourceID	Long Int.
LinearScale	Double

DRN

Field	Type
IJK	Long Int.
SPID	Long Int.
Elevation	Double
Cond	Double
IFACE	Long Int.
Condfact	Double
SourceID	Long Int.
LinearScale	Double

GHB

Field	Type
IJK	Long Int.
SPID	Long Int.
Bhead	Double
Cond	Double
IFACE	Long Int.
Condfact	Double
SourceID	Long Int.
LinearScale	Double

HFB6

Field	Type
IJK1	Long Int.
IJK2	Long Int.
Hydchr	Double
Factor	Double
SourceID	Long Int.

Figure 8.16 Tables associated with list-based MODFLOW stress packages.

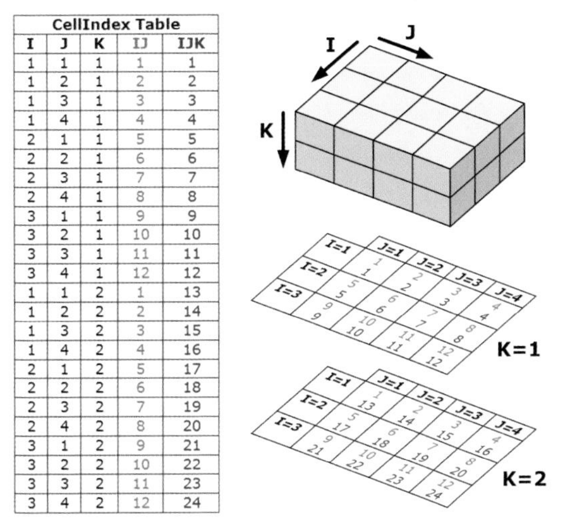
CellIndex Table

I	J	K	IJ	IJK
1	1	1	1	1
1	2	1	2	2
1	3	1	3	3
1	4	1	4	4
2	1	1	5	5
2	2	1	6	6
2	3	1	7	7
2	4	1	8	8
3	1	1	9	9
3	2	1	10	10
3	3	1	11	11
3	4	1	12	12
1	1	2	1	13
1	2	2	2	14
1	3	2	3	15
1	4	2	4	16
2	1	2	5	17
2	2	2	6	18
2	3	2	7	19
2	4	2	8	20
3	1	2	9	21
3	2	2	10	22
3	3	2	11	23
3	4	2	12	24

Figure 8.17 The CellIndex table is used to join tables and features in the MODFLOW data model.

the index by one as shown in figure 8.17. The IJ values are numbered similarly, but the numbering is repeated in each layer.

When used in conjunction with the MODFLOW data model, Cell2D and Node2D features include an IJ field, and Cell3D and Node3D features include an IJK field, where the IJ and IJK fields are populated using the strategy used in the CellIndex table (the IJ and IJK fields are equivalent to the CellID fields). For single-layer models or for 2D arrays (recharge, evapotranspiration, etc.), the MODFLOW tables can be joined directly to the Node2D and Cell2D features. For multilayer models and 3D arrays, the MODFLOW tables are first joined with the CellIndex table using the IJK field, and the result is then joined with Node2D or Cell2D features through the IJ field. For example, a map layer illustrating hydraulic conductivity values could be generated by first joining the IJK field of the LPFProperties table to the IJK field of the CellIndex table. The IJ field of the resulting temporary table would then be joined to the IJ field of the Cell2D feature class. The values associated with individual grid layers in the resulting map layer can be displayed using a definition query ("CellIndex.K = 1, CellIndex.K = 2, etc.). For transient models, a definition query on the SPID field can also be used to view the values associated with a particular stress period. Joins can be performed with Cell3D and Node3D features in a similar fashion.

Map layers of MODFLOW data associated with cell and node features can be symbolized and toggled on/off independently from other features in the table of contents (TOC) window in ArcMap and ArcScene. For example, data from a MODFLOW model of the Barton Spring region of the Edwards Aquifer was imported into the MODFLOW data model tables and by joining Cell2D and Node2D features with the tables a display of the MODFLOW data was created in ArcMap (figure 8.18).

MODFLOW Analyst

The MODFLOW data model provides the foundation upon which a rich suite of tools and utilities are constructed. The existence of a standard, the open nature of the data model, and the implementation of the data model in a fully scriptable GIS environment opens a new field of applications and research related to MODFLOW modeling. A suite of tools named "MODFLOW Analyst" has been developed as part of the Arc Hydro Groundwater tools. MODFLOW Analyst is used for building and managing MODFLOW models via the Arc Hydro Groundwater and MODFLOW data models. Most of the tools associated with MODFLOW Analyst are coded as ArcGIS geoprocessing (GP) tools (figure 8.19). This allows easy use from any ArcGIS application (ArcMap, ArcCatalog, ArcScene, or ArcGlobe). In addition, users can leverage the GP tools to create more advanced tools, or to combine a number of tools into models to support custom workflows.

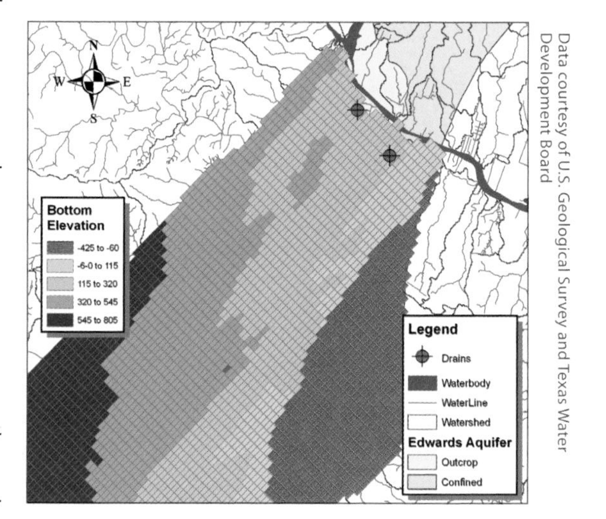

Data courtesy of U.S. Geological Survey and Texas Water Development Board

Figure 8.18 MODFLOW model of the Barton Spring segment of the Edwards Aquifer. Cell bottom elevations are displayed with Cell2D features, and springs modeled as "Drains" in MODFLOW are displayed with Node2D features.

MODFLOW Analyst includes the following sets of tools:

- Import/export tools: The Import MODFLOW Simulation tool imports a set of MODFLOW input files; builds the Boundary, Node2D and Cell2D features; and populates the appropriate tables in the MODFLOW data model. The tool uses a modified version of the MODFLOW code to read the model and then writes the features and tables using ArcObjects from within MODFLOW. Tools are also provided for exporting the components of a MODFLOW data model to individual MODFLOW input files.

- View tools: View tools create spatial views (map layers) of model inputs and results by joining tables in the MODFLOW data model with Cell2D and Node2D features. Simple filters can be used to view a selected MODFLOW grid layer or stress period and include only active cells in the query.

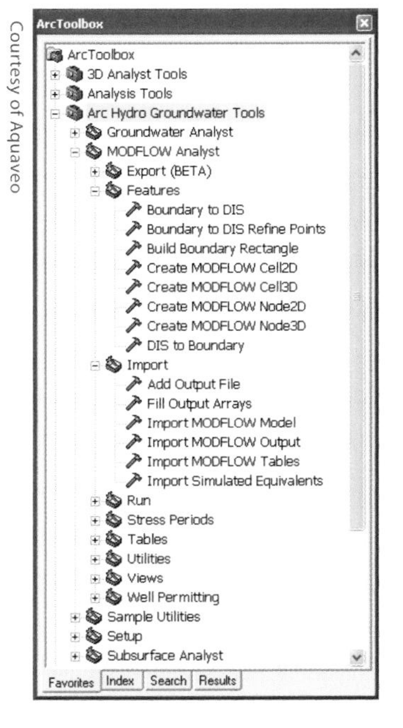

Courtesy of Aquaveo

Figure 8.19 MODFLOW Analyst geoprocessing tools (part of the Arc Hydro Groundwater tools).

- Building model input: One of the main benefits of the MODFLOW data model is that it facilitates the use of GIS features to construct model input. MODFLOW Analyst includes a suite of tools for populating the tables in the MODFLOW data model with new records using GIS features (points, lines, and polygons). These features are typically associated with the framework component of Arc Hydro, but the tools are developed in a generic fashion such that any set of features can be used. For example, a layer of polylines representing streams can be used to generate a new instance of the MODFLOW river package input. The high-level GP tools that build input data for a particular package are layered on top of low-level GP tools that overlay features with Cell2D polygons to find intersections, and low-level GP tools for intersecting generic time series data with stress periods.

- Custom workflows: With the MODFLOW Analyst tools as a foundation, it is possible to develop more advanced tools and custom workflows. For example, MODFLOW Analyst can be used for automated well permitting. This process consists of a few simple tools:

  1. Import MODFLOW Model: The MODFLOW model for the region in question is imported to ArcGIS. This model represents a calibrated steady-state model representing average current aquifer conditions.

  2. Establish Baseline Model: This tool adds a few supplemental tables to the system to manage the well-permitting process. It also takes simulated flows at selected rivers and water bodies representing baseline conditions and converts them to "observations" in the MODFLOW observations (OBS) process.

3. Create Well Feature: This tool is used to input the coordinates of a candidate well and create a new well in the Well feature class.

4. Create MODFLOW Well Records: This tool takes the new well and an associated discharge rate and creates a corresponding record in the Well Package table.

5. Analyze Well: This tool saves a copy of the modified well package and runs a new MODLFOW simulation. The resulting flows at selected rivers and water bodies are computed by the MODFLOW OBS process and are compared to baseline conditions. The changes resulting from the new well are calculated and tabulated in a report.

6. Reject Well: If a well is rejected, this tool is used to delete the candidate well from the database and restore the model to the baseline conditions.

7. Accept Well: If a well is accepted, this tool is used to convert the candidate well to a permanent well and to update the baseline condition of the model to include the new well.

In a similar fashion, workflows can be constructed for parameter estimation, Monte Carlo simulations, and risk analysis. Given that the data model is organized in a database, it is possible to define custom permissions for each user of the database. The data model can be stored on a central server, and Web services could access and manage the model remotely or via a Web interface.

Steps for implementing the Simulation component

This section outlines the main steps involved in populating the classes of the simulation component and integrating simulation models into the Arc Hydro Groundwater geodatabase. The first steps involved in implementing this component are to refine the geodatabase design to meet your specific project needs (see chapters 2 and 9 for a detailed description of these steps). For example, in the following case we use tables from the MODFLOW data model to store simulation outputs; these tables are not part of the core Arc Hydro Groundwater design but were added to complement the data model for a specific use, representing MODFLOW simulations. To start populating the simulation component you need to define the boundary of your model. You can then import/create Cell2D/Node2D and Cell3D/Node3D features. The features will be created based on the model discretization information (the number of layers, rows, and columns, and the grid spacing). Once your model grid/mesh is defined you can use custom tools to import/create model data or to export native model input files from GIS features. In the case of MODFLOW, the MODFLOW Analyst tools automate the process of creating the features, tables, and map layers. Finally, you can use the results of this process to create products such as maps, scenes, and reports. In addition you can develop custom workflows to support modeling related tasks. The following checklist provides a summary of the main steps for implementing the simulation component.

Checklist

1. Create the classes of the simulation component (manually using ArcCatalog or by importing from an XML schema)
2. Add project specific classes, attributes, relationships, and domains as necessary
3. Document datasets and changes made to the data model
4. Define the boundary of your model in the Boundary feature class
5. Import or create model grid/mesh features (Cell2D, Cell3D, Node2D, and Node3D)
6. Apply tools to:
   1) Import model data into the geodatabase
   2) Populate/edit model data using ArcGIS standard tools
   3) Export model data to native model input files
7. Visualize model data and create products (maps, scenes, reports)
8. Develop custom workflows

The following example illustrates the representation of a simulation model using the classes of the simulation component. As part of the Texas Water Development Boards' Groundwater Availability Modeling (GAM) program, three MODFLOW models covering the Edwards Aquifer were developed for simulating future groundwater availability in the aquifer. In this example we focus on the model of the Barton Springs segment of the aquifer (Scanlon et al. 2001), highlighted in figure 8.20.

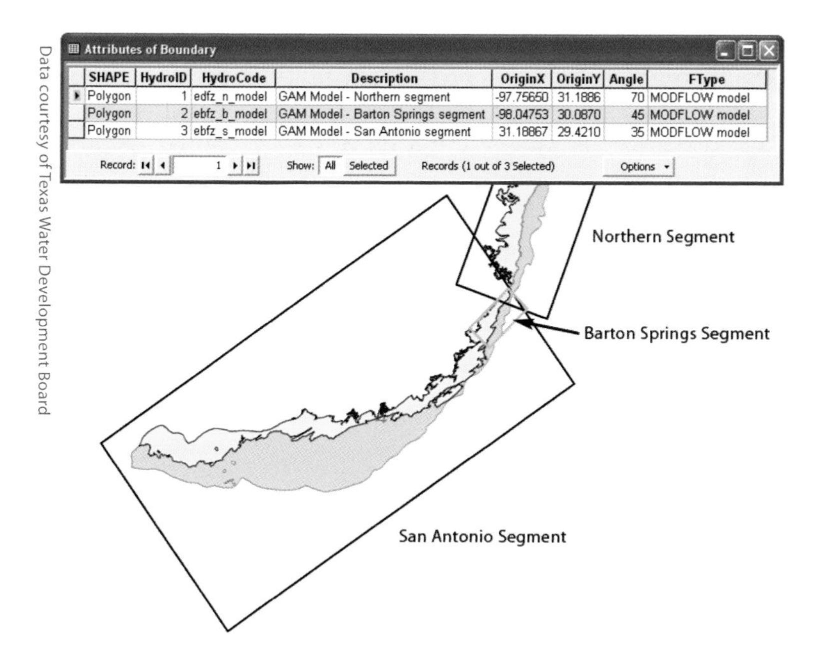

Attributes of Boundary

SHAPE	HydroID	HydroCode	Description	OriginX	OriginY	Angle	FType
Polygon	1	edfz_n_model	GAM Model - Northern segment	-97.75650	31.1886	70	MODFLOW model
Polygon	2	ebfz_b_model	GAM Model - Barton Springs segment	-98.04753	30.0870	45	MODFLOW model
Polygon	3	ebfz_s_model	GAM Model - San Antonio segment	31.18867	29.4210	35	MODFLOW model

Record: 1 Show: All Selected Records (1 out of 3 Selected) Options

Figure 8.20 The spatial location and orientation of a simulation model are defined by the model boundary. Highlighted is the boundary of the Barton Springs groundwater availability model.

The Import MODFLOW Model tool (part of the MODFLOW Analyst tools) was used to import the Barton Spring model. Inputs to the tool include the X and Y origin of the model and its angle of rotation, which define the location of the model in world coordinate space (figure 8.21). The length, width, and depth of the model are defined by the grid geometry as described in the discretization (DIS) portion of the MODFLOW input files.

Figure 8.22 shows the Cell2D and Node2D features created from the discretization information. Node2D features are located at the center of Cell2D features. The origin of the model is at the upper left corner of the boundary, thus the top left cell will be the first cell of the model grid (I = 1, J = 1, K = 1) and is assigned an index IJ = 1. The MODFLOW data model uses an IJ field that is equivalent to the more generic CellID and NodeID indices defined by Arc Hydro Groundwater.

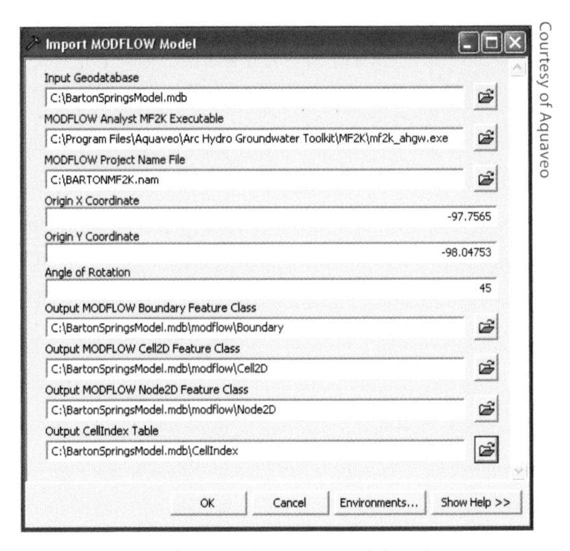

Figure 8.21 The Import MODFLOW Model tool imports a complete MODFLOW model into the geodatabase and creates Boundary, Cell2D, and Node2D features to represent the model grid.

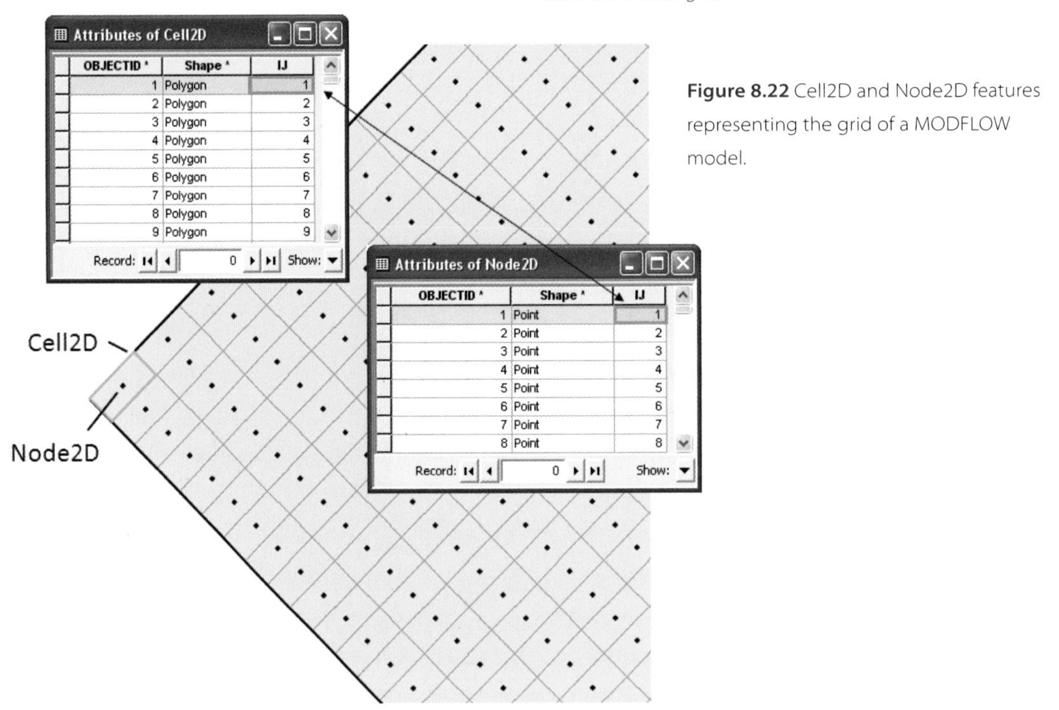

Figure 8.22 Cell2D and Node2D features representing the grid of a MODFLOW model.

The top and bottom elevations of the model cells can be imported from the discretization (DIS) file and read into the TopElev and BotmElev tables of the MODFLOW data model. Based on these elevations and the Node2D and Cell2D features, Node3D and Cell3D features can be created to visualize the model in ArcScene. Figure 8.23 shows an example of Cell3D features representing a 3D MODFLOW model grid rendered in ArcScene.

In addition to viewing the model grid and inputs, one can also run the model and visualize the results. The Import MODFLOW Output tool (part of the MODFLOW Analyst tools) was used to import the MODFLOW outputs (heads, drawdown, flow) into the appropriate tables of the MODFLOW data model. Figure 8.24 shows simulated heads imported into the OutputHead table of the MODFLOW data model. Each head value is indexed with an IJK attribute relating the computed head value with a MODFLOW cell, and the head values are also indexed with a TimeID attribute giving the time step of the model.

Attributes of OutputHead

Object ID	TimeID	IJK	Head
320	1	320	670.84
321	1	321	662.24
322	1	322	650.6
323	1	323	639.53
324	1	324	628.96
325	1	325	620.03
326	1	326	612.29
327	1	327	605.19
328	1	328	599.54
329	1	329	595.95
330	1	330	594.4
331	1	331	595.49
332	1	332	600.3
333	1	333	608.31
334	1	334	619.86
335	1	335	629.05
336	1	336	-9999

Record: 1 Show:

Figure 8.24 Simulated heads stored in the OutputHead table of the MODFLOW data model.

Courtesy of Michelle Smilowitz, Aquaveo

Figure 8.23 Cell3D features representing a 3D MODFLOW model grid rendered in ArcScene. Each cell in the model is represented by a Cell3D feature. Cells are symbolized to show the different hydrogeologic units.

Once the model data is in the geodatabase, model inputs and outputs can be visualized and analyzed by joining the tabular data (e.g., simulated heads) with Cell2D, Node2D, Cell3D, and Node3D features and creating data layers representing model data and results. For example, a head map can be created by joining Cell2D features with the output head table. Filtering the new layer for a TimeID will result in a head map for the selected model time step. In addition, time series can be plotted for specific cells of interest, and animations can be created to display changes over time (figure 8.25).

References

Harbaugh, A. W., E. R. Banta, M. C. Hill, and M. G. McDonald. 2000. MODFLOW-2000, the U.S. Geological Survey modular ground-water model—User guide to modularization concepts and the Ground-Water Flow Process: U.S. Geological Survey Open-File Report 00–92.

Scanlon, B. R., R. E. Mace, Brian Smith, S. D. Hovorka, A. R. Dutton, and R. C. Reedy. 2001. Groundwater availability of the Barton Springs segment of the Edwards aquifer, Texas—numerical simulations through 2050: The University of Texas at Austin, Bureau of Economic Geology.

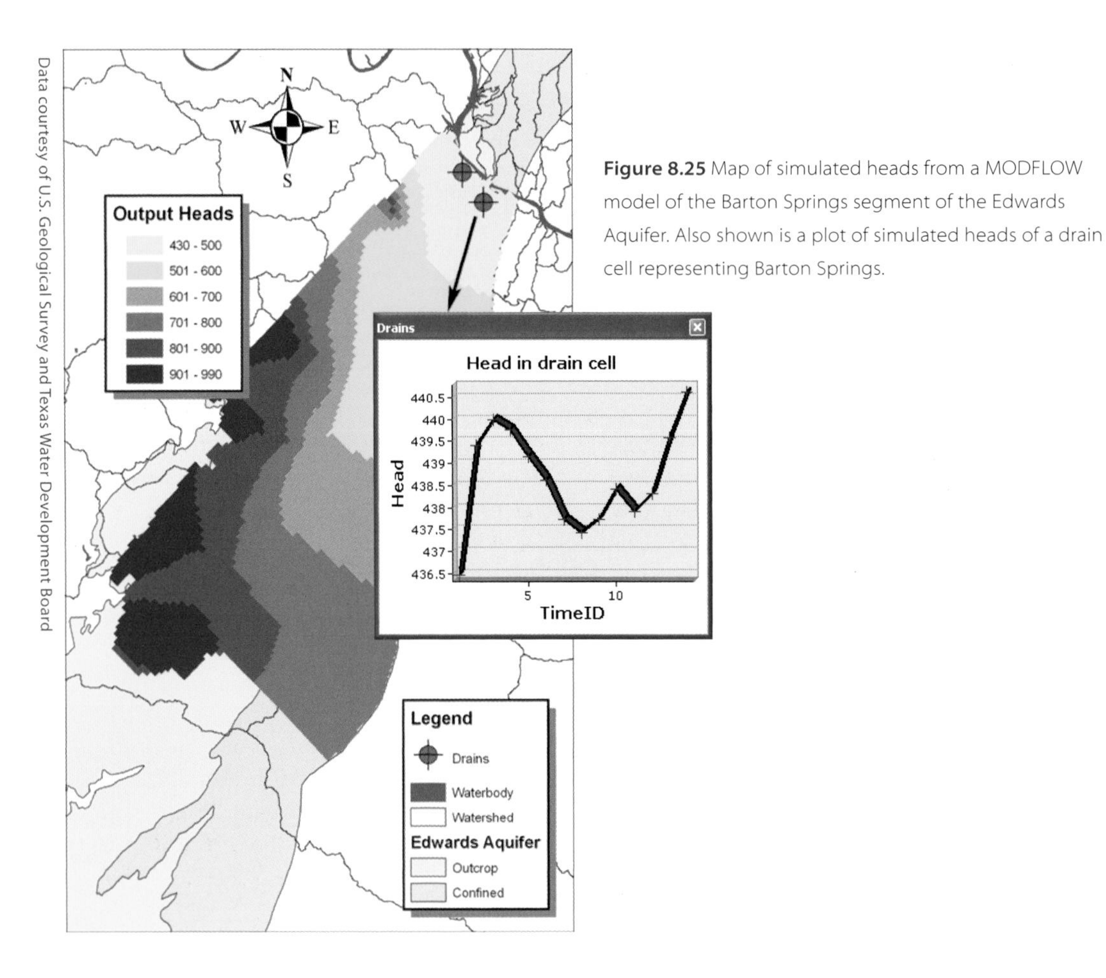

Data courtesy of U.S. Geological Survey and Texas Water Development Board

Figure 8.25 Map of simulated heads from a MODFLOW model of the Barton Springs segment of the Edwards Aquifer. Also shown is a plot of simulated heads of a drain cell representing Barton Springs.

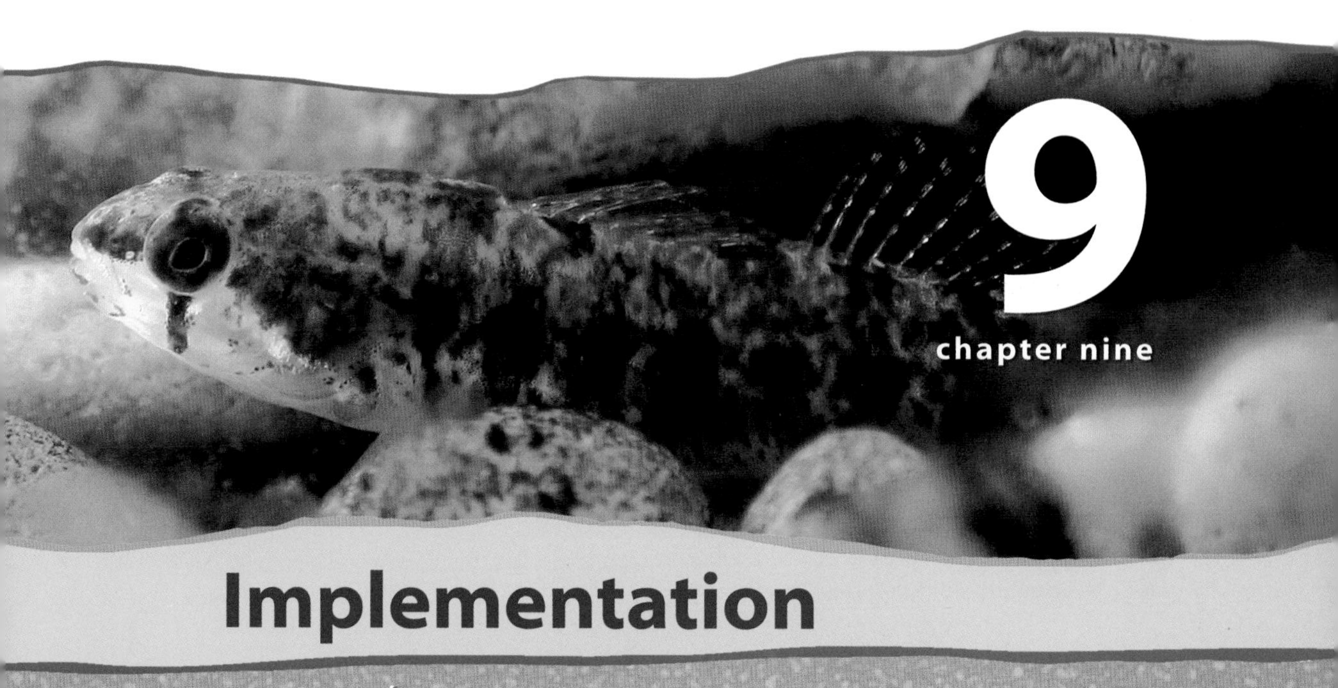

9

chapter nine

Implementation

STEVE GRISÉ AND GIL STRASSBERG

PREVIOUS CHAPTERS DISCUSSED THE CONTENTS OF THE ARC HYDRO GROUNDWATER data model and provided a number of examples. This chapter is designed to help you implement your own groundwater project and efficiently tailor the data model to fit your specific needs. Remember that the data model and associated materials are a starting point for your project rather than a final solution. Online resources for further information include `http://support.esri.com/` and `http://webhelp.esri.com`. The first step in this process is to download a design template, available at the Esri Web site (`http://resources.arcgis.com/ArcHydro`) and at the Arc Hydro Groundwater Web site (`www.archydrogw.com`).

Implementation process

Getting started on your groundwater project is straightforward. At a basic level, the process involves creating and refining a geodatabase and then documenting and deploying the design. For most projects, this simple approach is adequate. Toward the end of the chapter we also introduce some additional patterns and methods that should help you deal with more complex project requirements. You can use the following checklist as a guide for outlining tasks to be completed for your project.

Checklist

Status

1. Create an Arc Hydro Groundwater geodatabase
2. Refine data model content
3. Set spatial reference
4. Load data
5. Index features and build relationships, networks, and topologies
6. Set up tools and applications
7. Document
8. Deploy

Creating an Arc Hydro Groundwater geodatabase from a template

To create an Arc Hydro Groundwater geodatabase, you can use a template stored as an XML workspace document that contains the model. The XML document contains a complete description of a geodatabase (in this case an Arc Hydro Groundwater geodatabase) that can be imported to an existing geodatabase. This process will add the feature classes, tables, relationships, and domains defined in the XML document (figure 9.1). A detailed description on creating a new Arc Hydro Groundwater geodatabase is available on the Arc Hydro Groundwater Web site (`www.archydrogw.com`).

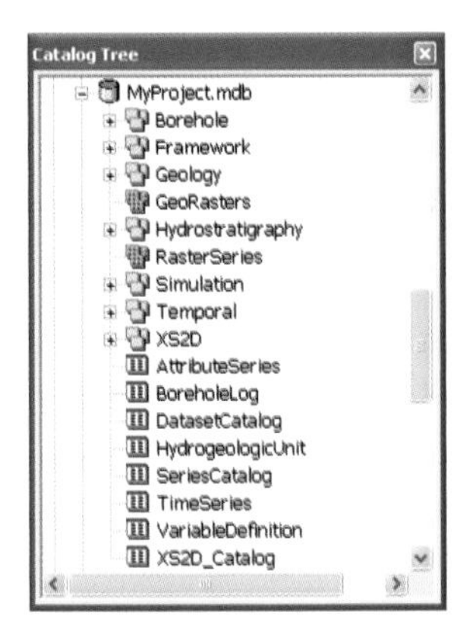

Figure 9.1 Example of an Arc Hydro Groundwater geodatabase created by importing the data model datasets from an XML document.

Customizing the data model

Your starting point for design is to review the template data model. Refining it for your own project involves adding (or removing) feature datasets, feature classes, fields, coded values, topologies, and any other necessary modifications. Most people spend time reviewing existing datasets in the original data model and making improvements so the new model better supports their specific data needs. Understanding what you need for a particular project typically involves reviewing existing datasets and talking with subject-matter experts and end users. You can use the conceptual groundwater data model for discussion purposes, and it's a good practice to make notes on a paper or digital copy of the model as you talk with your project team and users. Eventually, one person typically uses ArcGIS to make the model changes, and the number of changes is typically quite small (at least compared to the length of the discussions about changes!). It's relatively easy to get carried away in a complex series of meetings, and sometimes there is a tendency to overanalyze the design challenges. Thus, you can consider a few simple guidelines:

1. You, as the person leading this part of the project, need to own the design process and the design of the system. With that focus, you will likely learn a lot more by working with the data, maps, and tools required for your project than by talking to people.

2. Typically, the biggest gaps are in understanding how technology can be applied to solve longstanding problems. Take the opportunity to innovate during the design phase, because it gets harder to make changes once a new system is built. Prototype your ideas and refine the design based on practical problems that need to be solved. The ArcGIS Desktop Help system has a series of steps outlined in a section called

"Designing a geodatabase" that should help you in the process (`http://webhelp.esri.com`).

3. You should leverage the experience in your organization; somebody has probably thought about this problem a lot more than you have. You shouldn't pretend to be that person if you aren't, and you need to work closely with people who have the domain experience and expertise.

4. People will always tell you about the need for more data—information they don't have today that they believe is necessary for the organization. Be cautious about these requests. If it is really important to your business, the data already should be available. Otherwise, additional data acquisition and maintenance costs might be involved.

5. It's relatively easy to build a data model that your organization has no hope of managing on a day-to-day basis. Try to start with a simple implementable design with additional wish-list items that can be added after the initial implementation.

Using ArcCatalog to refine geodatabase contents

Like most Windows applications, you can use the "Properties…" context menu in ArcCatalog to modify the groundwater data model for your project. In the following example (figure 9.2), the properties of the Aquifer feature class are being inspected. In the Fields tab, a new text field called "Description" is added to the Aquifer feature class. This will allow users to add a detailed text description of the aquifer.

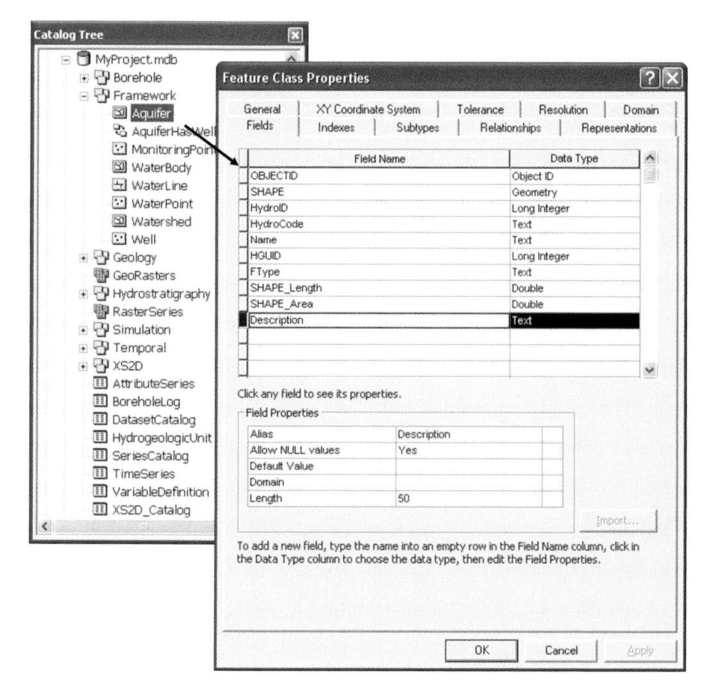

Figure 9.2. You can use ArcCatalog to customize the groundwater data model.

You can also customize the data model by adding lists of coded values specific for your project/organization. In the first Arc Hydro data model, FType fields were defined with coded value domains. These were intended for mapping/classification of different kinds of water features, so that we could distinguish between rivers and canals for mapping and analytic purposes. Historically, this was rooted in USGS hydrographic mapping. Many other organizations around the world involved in "blue-line mapping" had similar practices. In the first Arc Hydro data model we didn't provide any examples, but the USGS and other organizations developed fairly sophisticated implementations that in our experience were beyond what most project teams required. In the Arc Hydro Groundwater data model we included some sample domains for FType, and we kept the pattern quite simple. When you create the data model from the template XML document, a few of the feature classes (e.g., Well, Aquifer) will have coded values associated with the FType fields. These examples should form a good starting point for you to manage the types of features in your project, but it is likely you will need to edit the list or even add multiple attributes to achieve the results you are looking for.

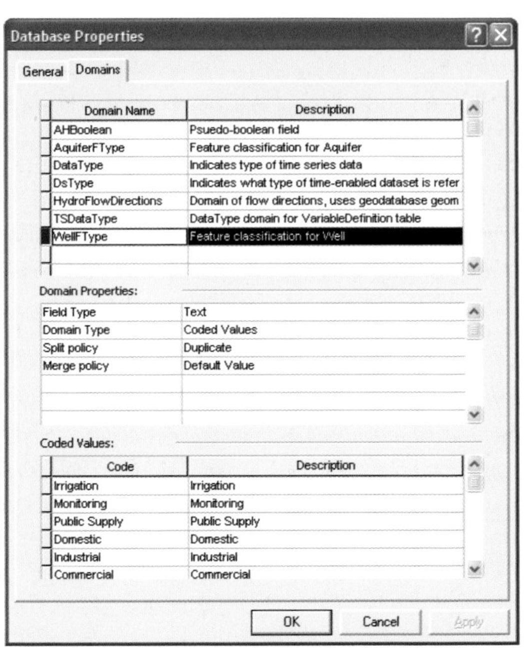

Figure 9.3 ArcCatalog interface for editing, adding, or deleting coded value domains. The upper section shows the domains available in the geodatabase, and the lower section shows the coded values defined for the selected domain.

Coded value domains are defined on the geodatabase level, which means that the same list of codes can be applied to multiple attributes in different feature classes and tables. You can modify coded value domains with Arc Catalog. When you select a domain name, the codes of the domain are shown in the bottom part of the interface. You can then modify, delete, or add values to the list. The following example (figure 9.3) shows a WellFType coded value domain created to define a set of unique values that can be entered in the FType field of the Well feature class.

After creating the coded value domain, we associate the FType field in the Well feature class with the WellFType Domain (figure 9.4). You can create as many domains as necessary for your project and assign those to fields within the geodatabase.

When a coded value domain is related to an attribute, the values in that attribute are limited to the coded values defined in the domain. For example, when you edit the FType field in ArcMap, values will be restricted to the set of coded values defined in the WellFType domain. In an edit session, a drop-down menu appears when you try to edit the FType values, and you must select one of the predefined coded values (figure 9.5).

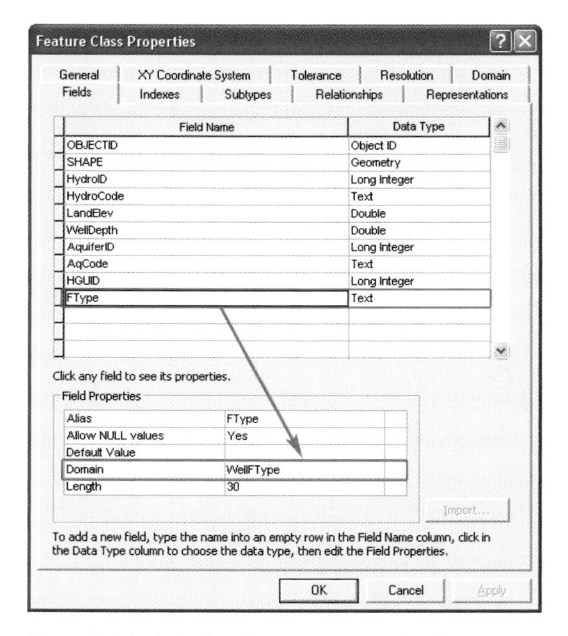

Figure 9.4 Coded value domain is assigned to the FType field in the Well feature class.

You can also consider using design tools available on the ArcGIS resources center (`http://resources.arcgis.com`), such as GDB XRay (search for "GDB XRay"), ArcGIS Diagrammer (search for "Diagrammer"), and CASE tools based on Unified Modeling Language (search for "Building geodatabases with CASE tools"). These tools can help to automate some of your design tasks, but for most projects it's easier to get started using ArcCatalog.

Set a spatial reference for datasets

An important part of the design process is the definition of spatial references that will be used in your project. The definition can be based on existing data available in your organization or based on planned datasets for which you will need to determine the appropriate spatial reference. It is important to get the spatial reference assigned correctly before loading data. While it is sometimes possible to change the spatial reference after loading the data, you should avoid this because it can cause errors while attempting to load data and result in shifts in geometry that may introduce inaccuracies in your data.

The spatial reference should be set at the feature dataset level when feature classes are contained within a feature dataset. This will set the

Attributes of Well

OBJECTID	Shape	HydroID	HydroCode	LandElev	WellDepth	AquiferID	AqCode	HGUID	FType
49	Point	49	5857204	800	245	114	218EDRDA	4	Stock
50	Point	50	5857205	829	360	114	218EDRDA	4	Unused
51	Point	51	5857206	940	400	114	218EDRDA	4	Domestic
52	Point	52	5857208	948	375	114	218EDRDA	4	Unused
53	Point	53	5857210	800	300	114	218EBFZA	4	Public Su
54	Point	54	5857301	883	312	113	218EBFZA	4	<Null>
55	Point	55	5857302	765	285	113	218EBFZA	4	Public Supply
56	Point	56	5857303	870	315	114	218EBFZA	4	Industrial
57	Point	57	5857304	780	380	114	218EBFZA	4	Irrigation
58	Point	58	5857305	809	415	113	218EBFZA	4	Domestic

Stock
Unused

Record: 53 Show: All Selected Records (1 out of 3613 Selecte ptior

Figure 9.5 A list of coded values appears when editing an attribute with a coded value domain.

properties for all feature classes in the feature dataset. If feature classes are outside a feature dataset, then you will need to specify each dataset separately. You can select a predefined spatial reference, modify it, or specify a new one. If you have existing data, you can choose to import the spatial reference from the existing data. Figure 9.6 shows an example of selecting the State Plane spatial reference from a list of spatial references that are supplied with ArcGIS.

In addition, for feature datasets containing 3D features (e.g., Borehole, Hydrostratigraphy, and Simulation), it is important to define the z coordinate system as well as the x and y coordinate system. Additional information on managing spatial references can be found in the ArcGIS Desktop help system (`http://webhelp.esri.com`). You can also use the Batch Project geoprocessing tool to batch project multiple datasets (figure 9.7).

The contents of the template data model are organized into different feature datasets in the geodatabase. Each component, such as framework

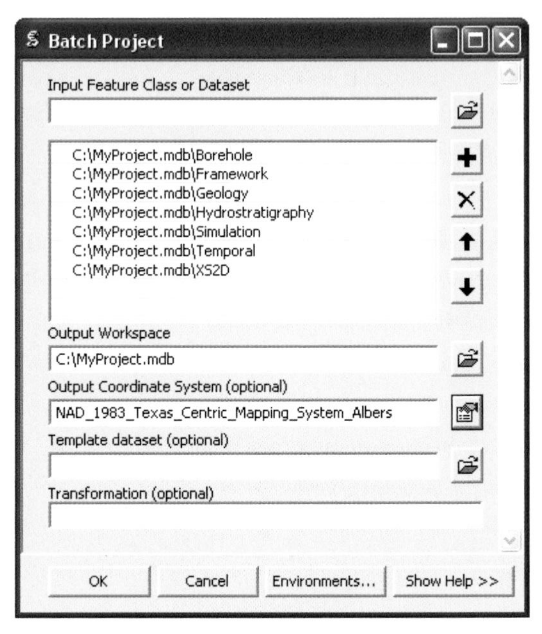

Figure 9.7 The Batch Project geoprocessing tool can be used to project the feature datasets of the groundwater data model to a selected coordinate system.

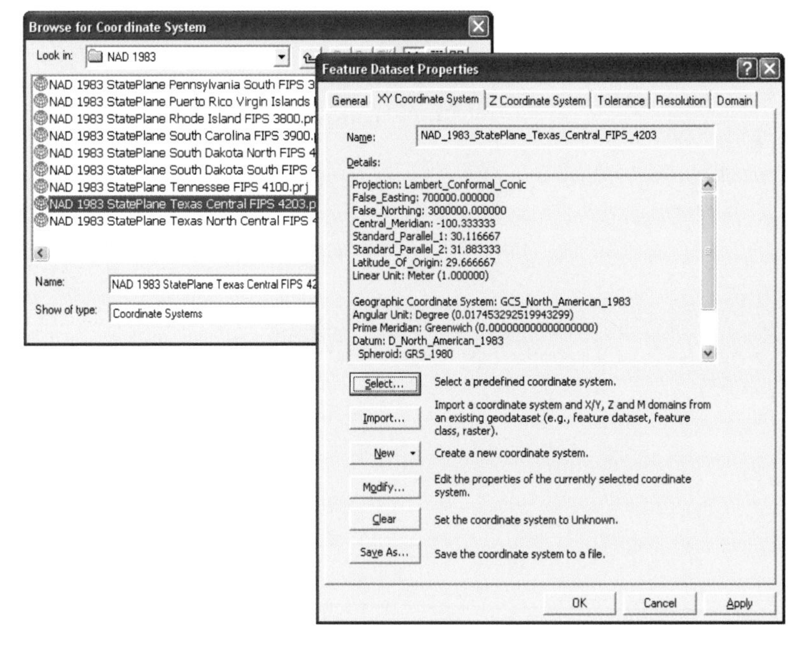

Figure 9.6 ArcGIS provides a list of spatial references to select from, and these can be customized as necessary.

or hydrostratigraphy, has its own feature dataset. This logical grouping easily can be changed if necessary, but you should be aware that all feature classes in a feature dataset must have the same spatial reference. Also, Esri recommends that all feature classes in the same feature dataset have the same permissions for editing and data management/access. If you have one workgroup editing the framework data and another workgroup managing the simulation datasets, you should plan the organization of the data according to editing privileges.

Loading data

This step involves using ArcCatalog data loading tools and possibly the ArcGIS Data Interoperability extension for more sophisticated data-loading tasks. Again, the ArcGIS Desktop Help system includes significant content to support data-loading tasks.

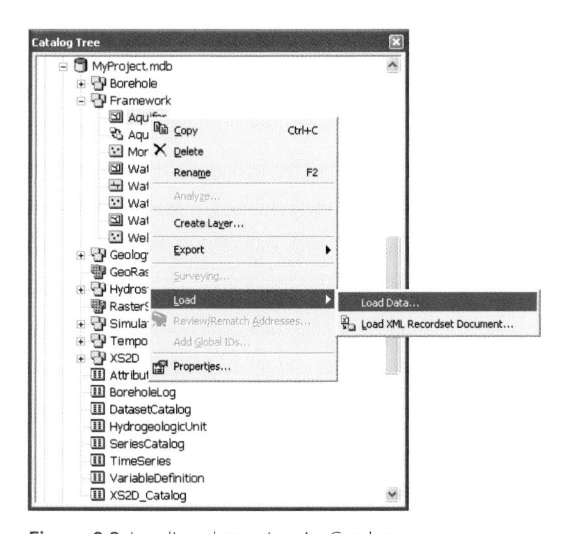

Figure 9.8. Loading data using ArcCatalog.

The simplest way to load data is to use the basic tools in ArcCatalog (figure 9.8).

There are also a number of geoprocessing and model-builder options for automating the data-loading process. You can use the Append geoprocessing tool to copy features and append them to your target feature classes (figure 9.9). The tool also allows you to match fields from your source datasets and the target feature classes.

The ArcGIS Data Interoperability extension is another useful toolset for loading data into the geodatabase. It provides a visual user interface for mapping source and target datasets. It also provides sophisticated extract, transform, and load (ETL) tools for data loading. The extension enables

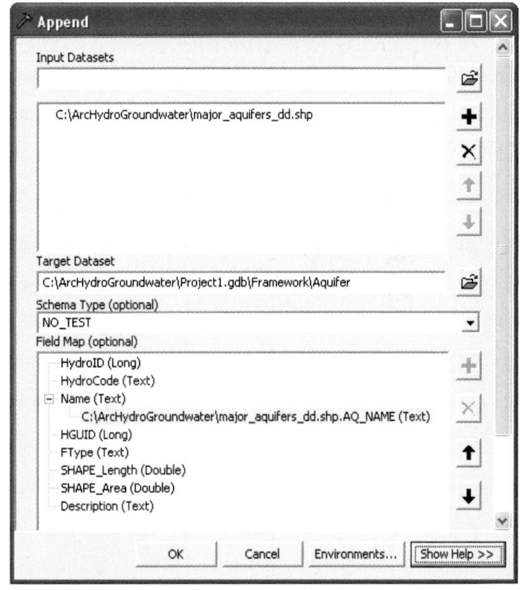

Figure 9.9 The Append geoprocessing tool can be used to load data from your source datasets to the target feature classes.

you to build loading workflows to automate repetitive loading tasks. Figure 9.10 shows an example of an ETL tool for loading well data from a text file. The tool automates the process of creating new Well features from the x and y coordinates and matching the fields in the source data with the fields in the target Well feature class. Again, the ArcGIS Desktop Help system provides documentation and examples that can help you efficiently load data into your data model.

Add geodatabase datasets such as networks and topologies

The groundwater data model does not require specific topologies, networks, annotation, or other datasets. You may choose, however, to add additional datasets to your project data model. For example, surface water networks can be created from the WaterLine and WaterBody feature classes. Another option is to add spatial topologies to define how spatial features interact. For example, you can define a topology rule that forces Well features to be inside an Aquifer feature (figure 9.11). Thus, this rule will be enforced when adding wells to your Well feature class.

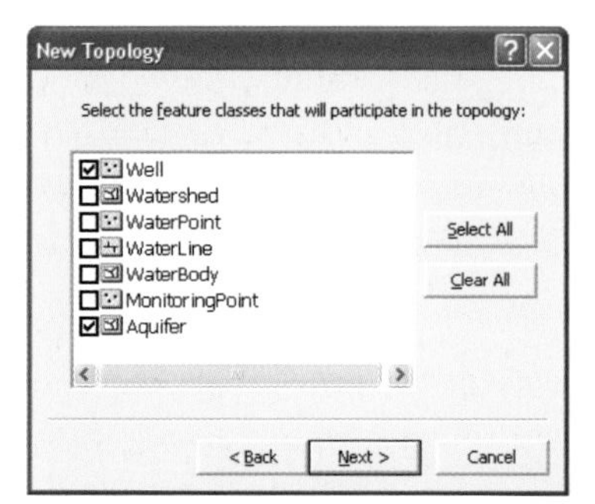

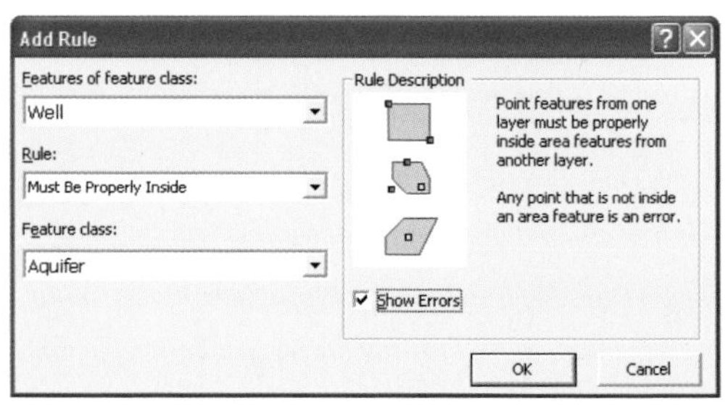

Figure 9.11 Example of a topology rule between Aquifer and Well feature classes.

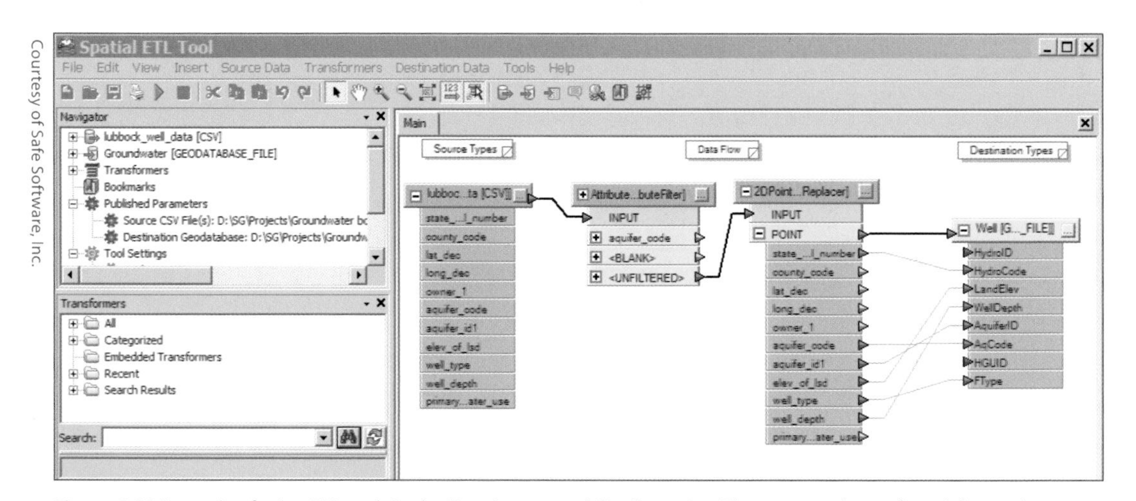

Courtesy of Safe Software, Inc.

Figure 9.10 Example of using ETL tools in the Data Interoperability Extension. The process shown from left to right: importing a text file, filtering by the aquifer code to get data for a specific aquifer, creating 2D points based on the x and y coordinates, and loading the points into the Well feature class.

Set up tools and applications

Install any tools and applications you plan to use and test with your data model. This will typically include basic editing and productivity tools and can also include more sophisticated modeling and simulation tools (e.g., the Arc Hydro Groundwater tools). It is also important at this stage to create some of the output products required for your project. If you've done a thorough job, you won't have any surprises here, but data models always require minor changes when you try to produce outputs for the first time. Make sure you do this before documenting and deploying the system.

Configure maps and services

While setting up your application environment, you can access Web services. Developing a plan for these services should be part of your implementation project, whether you need to access observations and measurements from partner systems or whether you simply need a base map in your Web-mapping applications. A simple example of using base-map services with your groundwater data is posted on the Arc Hydro Resource Center at `http://resources.arcgis.com/ArcHydro`. Examples on that site will guide you through setting up a dynamic map service and consuming online map services (figure 9.12).

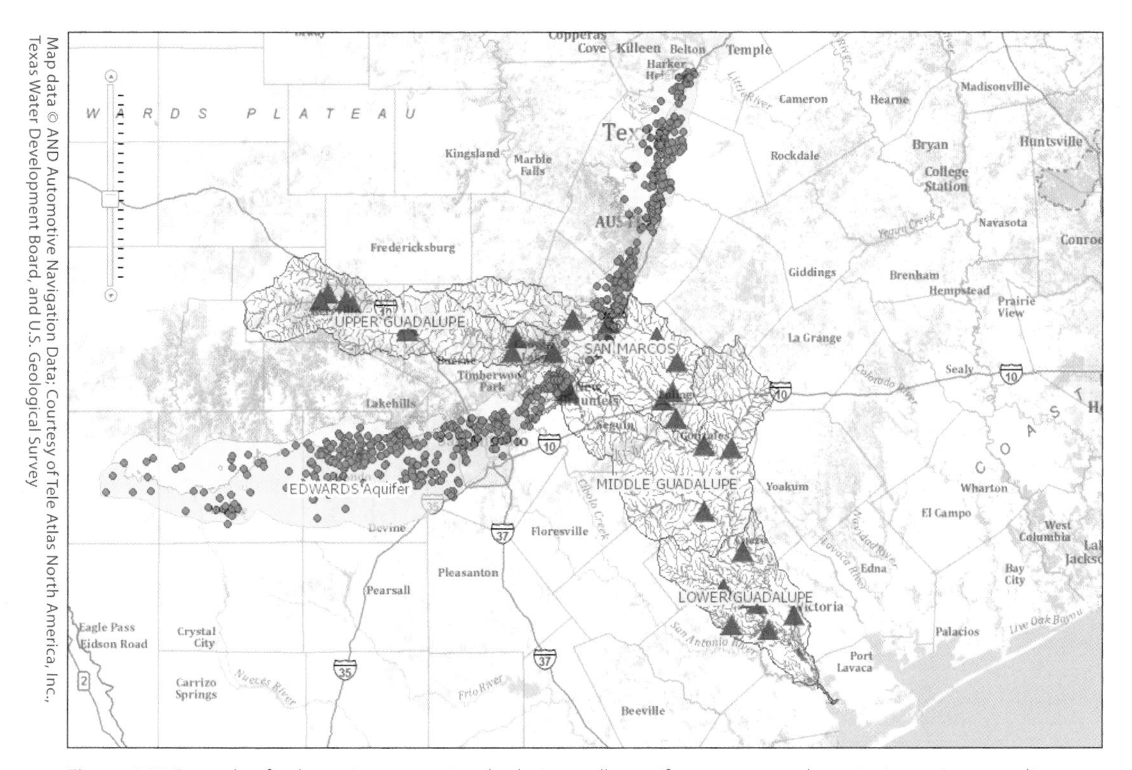

Map data © AND Automotive Navigation Data; Courtesy of Tele Atlas North America, Inc., Texas Water Development Board, and U.S. Geological Survey

Figure 9.12 Example of a dynamic map service displaying wells, aquifers, streams, and monitoring points stored in an Arc Hydro Groundwater geodatabase on top of a topographic base map published as a Web service by Esri.

You also have the ability to publish your own map and data services for other users to consume (figure 9.13). The example above demonstrates publishing ArcGIS Server map services, but other options to publish KML, Web Feature Services (WFS), and Geodata services from ArcGIS Server are also available.

You have different options to consume Web services, including ArcGIS applications and several Web mapping APIs. One example of using Web services is the consumption of Web Feature Service, which involves creating a data interoperability connection using ArcGIS Desktop (figure 9.14). Once a connection has been created, you can access the Web services using ArcMap like you would use other local datasets. While the performance of these services is not as fast as cached and dynamic map services, the one advantage of WFS in ArcMap is that you can work with these map layers as vector datasets and perform a number of operations like select, copy, paste, export, and other more advanced tasks.

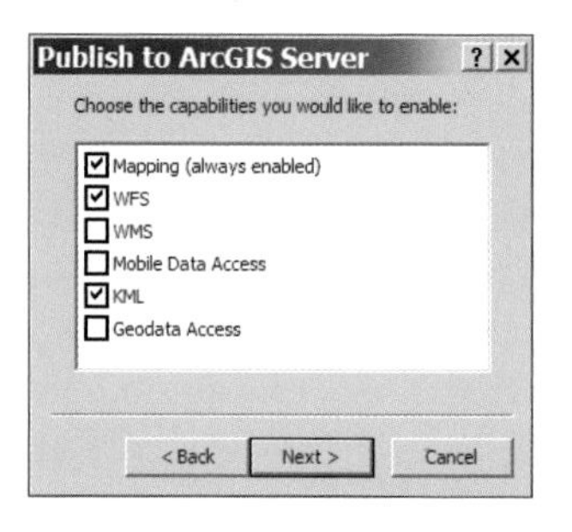

Figure 9.13 You can provide different capabilities for applications/users consuming your Web services.

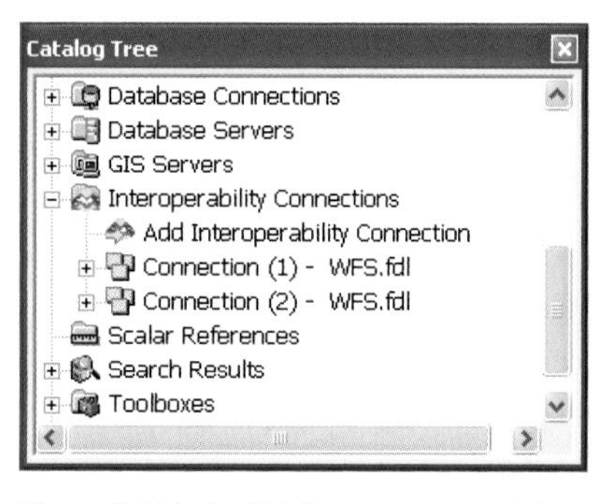

Figure 9.14 In ArcCatalog you can create interoperability connections to many types of services and consume them in ArcGIS Desktop.

Document the data model

You can document your data model several ways. The most basic way is to populate the metadata for each of the datasets in your geodatabase. You can use ArcCatalog to create standard metadata, which includes a description of the datasets, the fields in each dataset, and information on the spatial reference and extent of the data (figure 9.15). Some

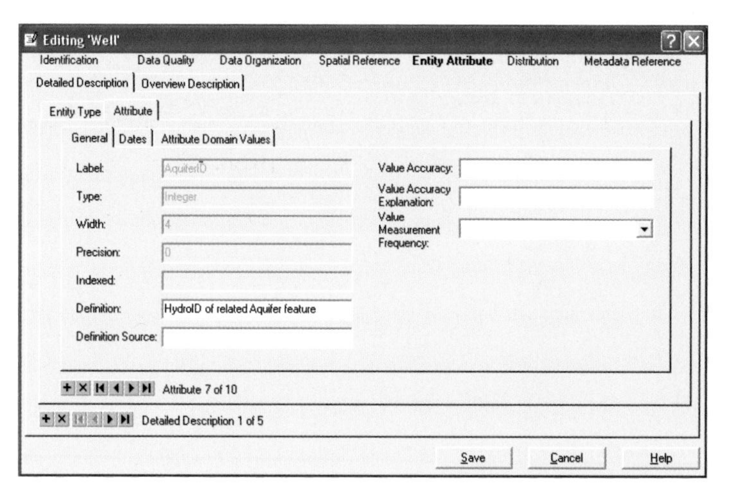

Figure 9.15 You can use ArcCatalog to create metadata describing the datasets within your geodatabase.

of the information is automatically populated for you, but you must add additional information to complete the documentation.

To document a more general view of your data model, you can use the version of the conceptual data model you developed with your project team. This approach works well for making a poster and documenting your design in a simple graphical way. You can find several tools to generate documentation at `http://resources.arcgis.com`. ArcGIS Diagrammer is a popular application for managing the design of your geodatabase using an XML Workspace document like ArcHydro-Groundwater.xml. The application also allows you to create visual diagrams that are good for documenting and sharing your design (figure 9.16).

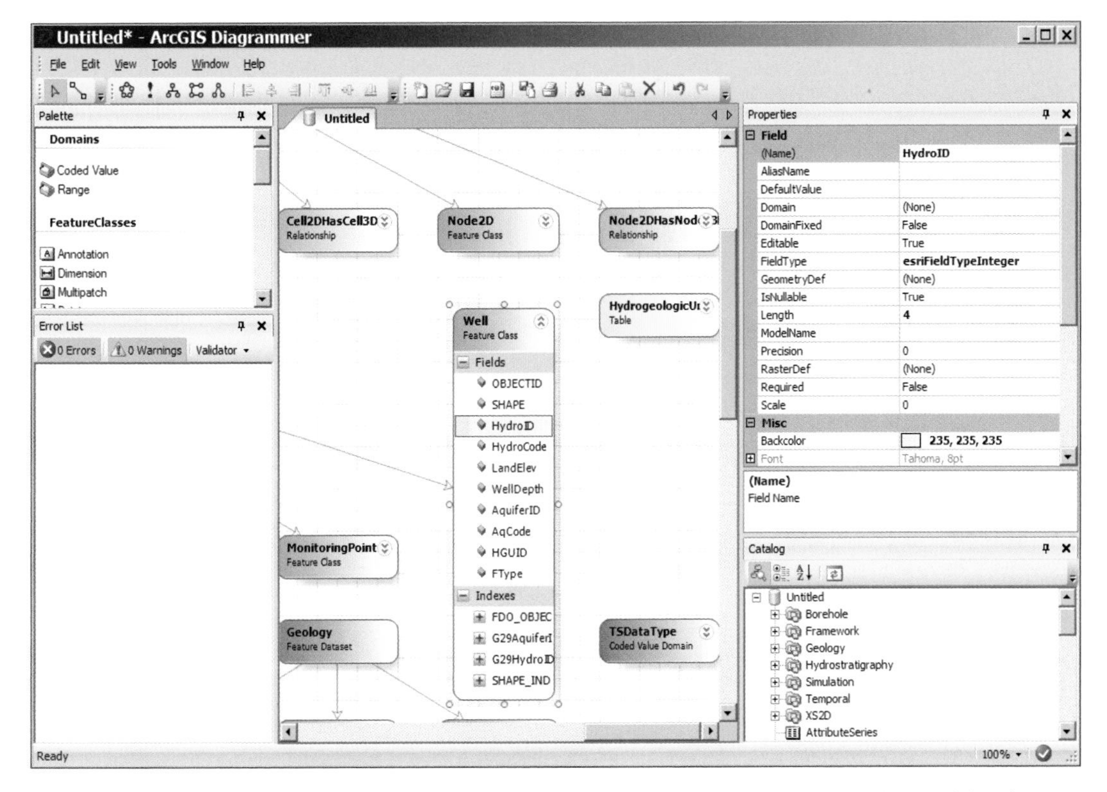

Figure 9.16 ArcGIS Diagrammer provides a graphical interface for editing and documenting your data model. In this example the properties of the Well feature class are inspected.

GDB XRay is another application that provides the ability to generate simple vector graphics (SVG) and data dictionary reports (figure 9.17). Many of the diagrams in this book were created with GDB XRay then further refined using Microsoft Visio.

Deploy to users

The best way to deploy the system to your users depends on the size of your user base. In many smaller projects, you are deploying to yourself and a few colleagues, so the challenges of testing, training, and deployment are minor. Larger workgroups and enterprise implementations require careful planning and testing to ensure that the system works correctly in a production environment. You can deploy these types of systems using many different organizational and IT practices. Most project teams vary in the way they test and deploy solutions to end users. In general, you should use functional and performance testing as necessary to prove that the system will work before deployment. As the number of users and applications increase, you will need to do more testing and validation of your solution. You should also develop training and technical support plans that suit the scale of your deployment. A suitable change management plan for the system should also be in place prior to going live with the system.

Design patterns and tradeoffs

In previous chapters you have seen solutions to a number of design challenges. Each of these solutions took some time to develop, and there are always tradeoffs in any solution. The next section describes some of the key design patterns in more detail to help you understand some of the thinking that went into the data model.

Well	A point that represents the location of a well and associated attributes
HydroID	Unique feature identifier in the Geodatabase
HydroCode	Permanent public identifier of the feature
LandElev	Land surface elevation or reference elevation
WellDepth	Depth of feature
AquiferID	HydroID of related Aquifer
AqCode	Text description for the aquifer related to the well
HGUID	Identifier of the hydrogeologic unit
FType	Classification of the Feature Type for mapping and analytic purposes

Well - FeatureClass

Name	Well
ShapeType	Point
FeatureType	Simple
AliasName	Wells
HasM	false
HasZ	false
Description	A point that represents the location of a well and associated attributes
DataTheme	ArcHydroFramework

Field	DataType	Length	AliasName	Description	Domain	DefaultValue	IsNullable
HydroID	Integer	4	HydroID	Unique feature identifier in the Geodatabase			true
HydroCode	String	30	HydroCode	Permanent public identifier of the feature			true
LandElev	Double	8	LandElev	Land surface elevation or reference elevation			true
WellDepth	Double	8	WellDepth	Depth of feature			true
AquiferID	Integer	4	AquiferID	HydroID of related Aquifer			true
AqCode	String	30	AqCode	Text description for the aquifer related to the well			true
HGUID	Integer	4	HGUID	Identifier of the hydrogeologic unit			true
FType	String	30	FType	Classification of the Feature Type for mapping and analytic purposes	WellFType		true

Figure 9.17 Examples of graphics and data dictionary reports created using GDB XRay.

Representing wells and boreholes

We took a simple approach in representing wells and boreholes in the data model. Wells are point features, and they can be associated with BoreholeLog data from which 3D spatial features (BoreLines and BorePoints) can be created. It is also possible to consider wells as 3D line features and to set them up to have measures and z-values in their geometry. This would permit changing the BoreLine and BorePoint classes to tables and using ArcGIS Linear Referencing to display the borehole data along the path of the well. This alternate approach appears to be more suited for petroleum and other drilling activities (where most of the wells are not vertical) than for groundwater data, but you can consider this alternative, especially if your data are focused on displaying 3D information along nonvertical wells.

One-to-many versus many-to-many associations

The template data model includes a number of one-to-many (1:M) associations, such as the relationship between Aquifer and Well features and the relationship between Cell2D and Node2D features. With different implementation strategies these relationships can be substituted with many-to-many (M:N) relationships. For example, a well can be screened across multiple aquifers or can be associated with different aquifers depending on the scale of the project. To support such associations one can develop an M:N relationship between wells and aquifers. The following example (figure 9.18) shows a conceptual M:N relationship between Well and Aquifer features. An intermediate table (named Screen) is created to support the M:N association. In this example, two wells are highlighted: Well 1 and Well 2. Each of the wells has a HydroID and is related with two Aquifer features. For each well, two instances are created in the intermediate table, the WellID attribute in the intermediate table references the HydroID of the Well feature and the AquiferID references the HydroID of an Aquifer feature. Through the intermediate table we associate Well 1 with Aquifer 1 and Aquifer 2, and Well 2 is related with Aquifer 2 and Aquifer 3. One can also query in the opposite direction. For example, if we start from Aquifer 2 and query the intermediate table, we will see that we have two wells (Well 1 and Well 2) related to this aquifer. Thus, a many-to-many relationship is established because a well

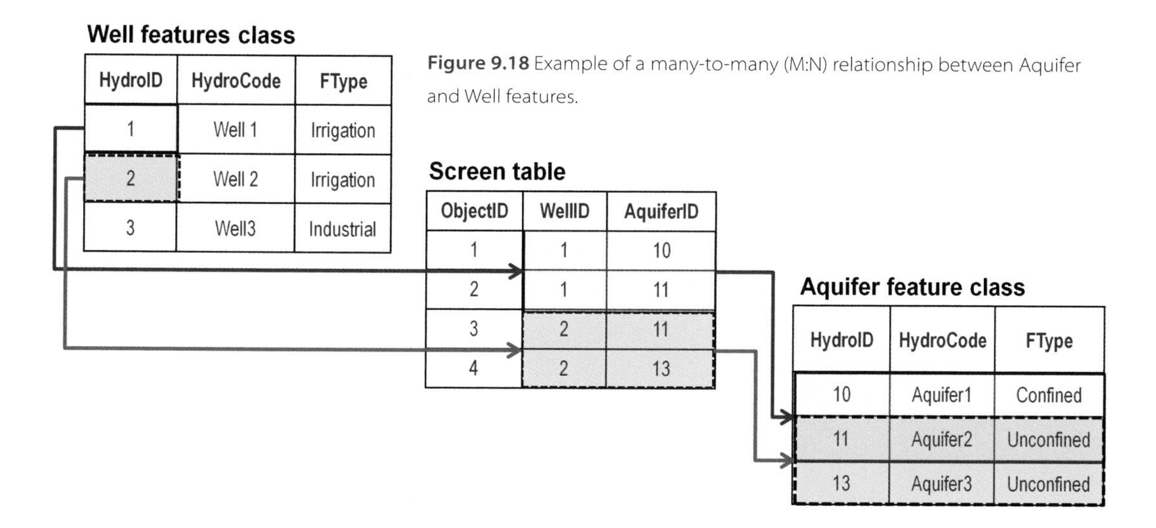

Figure 9.18 Example of a many-to-many (M:N) relationship between Aquifer and Well features.

can be related to multiple aquifers, and an aquifer can be related to multiple wells.

While it is possible to implement M:N associations, experience shows that managing data in M:N associations is more difficult from a user perspective. For this reason, we implemented a 1:M association in the template model. You should consider implementing a M:N relationship if necessary in your project. Another approach is to create additional relationships between Well and Aquifer features (see chapter 5 for a more detailed description of this approach).

Beyond the template—additional design methods

The template data model is not always well suited for your needs. For example, you might need additional content that does not exist in the Arc Hydro Groundwater template data model, or your data needs may be quite different from what the model provides. In general, the approach to tackling these types of challenges is similar to the process we went through while developing the groundwater data model. For a more complex project you can use some of the same techniques to add content for your needs. This involves a combination of database and GIS design methods:

- Start with the template conceptual data model.
- Work with subject-matter experts and your project team to identify data needs and refine the conceptual model.
- Look at your information products: maps, tools, Web services, and other requirements to determine the information needs of your users.
- Refine the conceptual data model to support the creation of maps, layers, Web services, and other content.
- Test and refine with real data. Get feedback early—people may not be able to provide much help on design details, but they will provide feedback when you show them data, maps, and tools.
- Build logical and physical models to support the design. Work as a single team with people who are building the applications and infrastructure for the project.
- Document and deploy.

Afterword

IN THIS BOOK WE PROVIDED A DESCRIPTION OF THE ARC HYDRO GROUNDWATER data model, including the design concepts, data structures, and the common uses one may find for the data model. The chapters of the book describe the different components of the data model and include an implementation section in each chapter to help you get started with applying the data model to your own projects. We hope that the book serves as a starting point for getting acquainted with the data model so that you can leverage the presented knowledge in your own projects.

Remember that the Arc Hydro data model, introduced in 2002 with the publication of *Arc Hydro: GIS for Water Resources*, is not a fixed design. Like water, the data model is continuously adapting to a fast-changing technological environment. The Arc Hydro Groundwater data model is a natural outgrowth of that earlier work. By the time you read this book, the evolving field of GIS will have introduced new technologies, providing new opportunities for us to improve the way we integrate GIS and water resources. Emerging fields of GIS will greatly affect the way we use the technology for hydrologic analysis. These include: 1) the development of time-aware spatial datasets, 2) the improvement of 3D datasets and tools, 3) the emergence of ArcGIS Online and the adoption of Web services by many agencies to serve spatial and temporal data over the Web, and 4) the evolution of GeoDesign to enable wider communities to quickly visualize and test their designs within a geographic context. These new technologies, and others we cannot foresee, will greatly impact the Arc Hydro data model and its tools and how they are used in your projects.

This book does not stand on its own; it is part of a wider set of resources that include papers, presentations, case studies, tools, and Web sites with updated designs. We encourage you to visit these Web sites and even provide your own content:

- A community-based Wiki where you can add your own content about the groundwater data model: `www.archydrogw.com`.
- The Arc Hydro Resource Center Web pages: `http://resources.arcgis.com/archydro`

Equally important to the geodatabase design presented in this book are tools that support the implementation of the data model. As mentioned in the book, Esri and Aquaveo have developed a set of tools named Arc Hydro Groundwater Tools to support the implementation of the groundwater data model: `http://www.aquaveo.com/archydro`.

From its initial conceptualization, Arc Hydro has tried to serve as a bridge between two communities: the first dealing with water resources and hydrology and the second dealing with GIS. We hope water resource engineers, hydrologists; hydrogeologists, geographers, and GIS specialists will all find this book a valuable addition to their respective fields. We believe educators will also find *Arc Hydro Groundwater: GIS for Hydrogeology* a valuable academic resource. Since its publication, *Arc Hydro: GIS for Water Resources* has been used as a textbook, not only for teaching GIS concepts but scientific concepts as well. The ability to show hydrologic concepts within a map is a strong learning method, as students can visualize hydrologic theory at a real location and time and with a true geographic context. We anticipate that educators will adopt *Arc Hydro Groundwater* as a textbook for college and postgraduate classes dealing with GIS and hydrogeology.

As you find new ways to apply the groundwater data model and the concepts contained in this book, we invite you to share your insights with other community members to help advance the use of GIS within the field of water resources and hydrogeology.

Glossary

APUNIQUEID Table for tracking HydroID values. Each time a new HydroID is assigned to a feature in the geodatabase, a counter is updated so that the same HydroID is never assigned again within the given geodatabase.

aquifer Group of formations, a single geologic formation, or part of a formation that contains sufficient saturated permeable material to yield significant quantities of water to springs and wells.

Aquifer (feature class) Polygon feature class for representing data from aquifer maps. Each Aquifer feature in the feature class represents the areal extent of an aquifer or a region within an aquifer.

aquifer map Common data product that portrays the boundary of aquifers over a given region and can also distinguish between zones within an aquifer.

Arc Hydro The overall data model for representing hydrology, including surface water and groundwater. Within this book we refer to this general model as the Arc Hydro data model, or simply as Arc Hydro.

Arc Hydro Framework Data model for representing the basic surface water and groundwater features. Within this book we refer to the framework data model as the Arc Hydro Framework, or simply as the framework.

Arc Hydro Groundwater The groundwater components of the Arc Hydro data model, which is the focus of *Arc Hydro Groundwater: GIS for Hydrogeology*. The book often refers to this data model simply as the groundwater data model.

Arc Hydro Groundwater Tools ArcGIS tools, for groundwater analysis, designed on top of the Arc Hydro Groundwater data model. The tools are developed collaboratively by Aquaveo and Esri and are available on the Aquaveo Web site (`www.aquaveo.com/archydro`).

Arc Hydro Tools ArcGIS tools, for surface water analysis, designed on top of the Arc Hydro data model. The tools are developed by Esri and available on the Arc Hydro Resource Center (`www.arcgis.com/archydro`).

AttributeSeries Table for archiving time-series data where multiple variables are indexed with the same feature and time.

borehole A hole drilled into the subsurface.

BoreholeLog Table for representing vertical data along boreholes. The table is the basis for creating 3D features to represent vertical borehole data as 3D geometries.

BorePoint and BoreLine 3D (z-enabled) point and line feature classes that represent point and interval data along boreholes.

Boundary (feature class) Polygon feature class that represents the 2D extent and orientation of a simulation model.

Cell2D Polygon feature class that represents cells or elements associated with a 2D simulation model or a single layer of a 3D model.

Cell3D Multipatch feature class that represents 3D cells and elements of a simulation model.

CellIndex Table containing indexes of MODFLOW cells and nodes used for joining tables and features in the MODFLOW data model.

data model Framework for describing a subject and storing data about it. Data models help us describe systems using structured sets of data objects.

DatasetCatalog Table for indexing time-series datasets within a geodatabase. DatasetCatalog is typically associated with RasterSeries or FeatureSeries in the geodatabase.

feature A row in a feature class representing a spatial object, including its geometry and attributes.

feature class A collection of features with the same geometry type, attributes, and relationships. A feature class is a table in the geodatabase that contains a custom field (Shape) for defining the geometry of features (point, line, polygon, multipoint, and multipatch).

feature dataset Container for a collection of spatially related feature classes together with relationship, topology, and network classes. The feature dataset is used for grouping feature classes either by spatial reference (all features classes in a feature dataset share the same spatial reference) or thematically.

FeatureSeries Collections of features indexed by time representing a series of geometries varying in location or shape. Each feature in a feature series exists for only a period of time. Features in a feature series can be grouped to form a "track" representing the change (in geometry, location, or both) of a particular feature over time.

GeoArea Polygon feature class representing the 2D extent of hydrogeologic units.

geodatabase Repository of geographic information organized into geographic datasets within a relational database system. It provides a common data storage and management framework for storing datasets supported in ArcGIS.

geographic data model Data model for representing the real world (or part of it) expressed with spatial datasets within a GIS.

geologic map Cartographic product containing information about the kinds of earth materials in a specific geographic area, the boundaries that separate them, and the geologic structures that have deformed them.

geologic map database Digitally compiled collection of spatial (geographically referenced) and descriptive geologic information about a specific geographic area.

GeologyPoint, GeologyLine, and GeologyArea Feature classes for representing data from geologic maps.

geomodel A combination of geo-objects that provides an abstract digital representation of a part of the earth's subsurface.

geo-object Distinctive subsurface features with measurable spatial boundaries in 3D. Geo-objects describe discrete entities such as a rock layer, fault, or a volume element.

GeoRasters Raster catalog for storing and indexing raster datasets that describe properties of hydrogeologic units.

GeoSection Multipatch feature class representing 3D panels for constructing vertical cross sections. Each feature in the GeoSection feature class represents a slice of a hydrogeologic unit, and a set of GeoSection features represent a complete cross section across multiple units.

GeoVolume Multipatch feature class for representing hydrogeologic units as 3D volume objects.

groundwater simulation model Simplified mathematical representations of aquifer systems used to interpret the flow of water and transport of contaminants within an aquifer and to predict how these will change under future stresses.

GSIS Geoscientific Information Systems, also known as geomodeling systems. These information systems differ from traditional GIS in their capabilities to represent complex 3D objects using either surface representations or volume objects.

HGUID Hydrogeologic unit identifier. Index used to associate features with descriptions of hydrogeologic units defined in the HydrogeologicUnit table.

HorizonID Index defining the vertical arrangement of hydrogeologic units in a depositional sequence. HorizonID values are assigned from bottom to top such that the smallest HorizonID is given to the base unit (the first layer in a deposition sequence), and the largest is given to the top unit.

hydro feature Feature classes customized as part of the Arc Hydro data model by adding HydroID and HydroCode attributes to them.

HydroCode Text attribute that is a permanent public identifier of a feature. The HydroCode provides a linkage with external information systems.

hydrogeologic framework A geologic framework that defines a distinct hydrologic system. In a hydrogeologic framework, subsurface materials are classified not only by the rock properties but also by the hydraulic properties that effect water storage and flow.

hydrogeologic unit Any soil or rock unit or zone which by virtue of its hydraulic properties has a distinct influence on the storage or movement of groundwater.

HydrogeologicUnit Tabular representation of hydrogeologic units. Attributes associated with hydrogeologic units are defined in the table, and spatial features created to describe the spatial location and extent of the units are related back to the conceptual definition in the table.

HydroID An integer attribute that uniquely identifies objects in an Arc Hydro geodatabase. The HydroID differs from the ObjectID, as it is unique across the geodatabase and not only within a certain class (table, feature class, raster catalog).

interface data model A geodatabase design for storing the entire contents of a given simulation model.

LAYERKEYTABLE Table used to manage multiple identifiers, so that data from different sources or for different projects or separate areas can be attributed with separate sets of unique identifiers.

MODFLOW Modular finite-difference flow model developed by the USGS. The most widely used groundwater simulation model.

MODFLOW Analyst A suite of ArcGIS tools developed as part of the Arc Hydro Groundwater tools. MODFLOW Analyst is used for building and managing MODFLOW models via the Arc Hydro Groundwater and MODFLOW data models.

MODFLOW data model (MDM) An interface data model developed as an extension to the Arc Hydro Groundwater data model. The MDM contains a series of tables and relationships and supports the storage of an entire MODFLOW simulation within a geodatabase.

MonitoringPoint Point feature class for representing locations where water is measured.

multipatch Geometry composed of 3D rings and triangles that represents objects that occupy a 3D area or volume. These can be geometric objects such as a cube or a sphere, or represent real-world objects such as buildings or trees. In the groundwater data model, multipatches represent 3D features such as GeoVolume, GeoSection, and Cell3D features.

Node2D Point feature class that represents the computational nodes in a 2D simulation model or a single layer of a 3D model.

Node3D Z-enabled point feature class for representing the computational nodes of a 3D simulation model grid.

piezometric head map Map describing the spatial patterns of groundwater levels (or pressure) in aquifers.

raster catalog Container for storing, indexing, and attributing raster datasets.

raster dataset Represents imaged, sampled, or interpolated data on a uniform rectangular grid.

RasterSeries Raster catalog for storing collections of raster datasets indexed by time. Raster series are useful for describing the dynamics of spatially continuous phenomena, like the variations in groundwater levels or the distribution of rainfall over time.

relationship Data structure that defines the association between objects in two classes using common attribute values in key fields.

SectionLine Polyline feature class for representing 2D cross-section lines on a map.

SeriesCatalog Table for indexing and summarizing time series stored in the TimeSeries table.

stratigraphic units Rocks or bodies of strata recognized as a unit for description, mapping, or correlation purposes.

Subsurface Analyst A suite of ArcGIS tools developed as part of the Arc Hydro Groundwater tools. Subsurface Analyst is used for creating, editing, and managing 2D and 3D hydrogeologic data within ArcGIS.

table A dataset for storing data values as an array of rows and columns.

time series A sequence {v, t} that describes the values, v, of a variable indexed against time, t. The value of t is called the *time stamp*, which records an instant of time used to reference the value, v.

TimeSeries Table for storing single-variable time-series data. The table implements a 3D structure for storing time-series values indexed by location, time, and variable. Each row in the TimeSeries table represents a value of a particular variable at a particular time associated with a particular feature.

TsTime Attribute representing the time stamp specifying the date and time associated with a time-series value.

TsValue Attribute containing a numerical value of a variable at a given location and time.

UTCOffset Attribute representing the number of hours the time coordinate system used to define TsTime is displaced from Coordinated Universal Time.

VariableDefinition Table for defining temporal variables. Each variable defined in the VariableDefinition table is uniquely indexed with a HydroID, and space-time datasets are related to the variable definition by referencing the HydroID of the variable defined in the table.

VarID Attribute containing a numerical identifier of a variable. Matches to the HydroID of a variable defined in the VariableDefinition table.

VarKey Unique text identifier for a variable, used when a variable is indexed in an attribute series table via field names.

Waterbody (feature class) Polygon feature class for representing areal water features in the water system, such as lakes, ponds, swamps, and estuaries.

WaterLine Line feature class for representing hydrographic "blue lines," which represent mapped streams and water body center lines.

WaterPoint Point feature class for representing hydrographic features such as springs, water withdrawal/discharge locations, and structures.

Watershed (feature class) Polygon feature class for representing drainage areas contributing flow from the land surface to the water system.

Web service Internet-based program built to provide a particular set of functionality or services.

well Human-made excavation or structure created in the ground to access groundwater for water extraction, injection, or monitoring.

Well (feature class) Point feature class for representing well locations and basic well attributes for identification, 3D representation, and linkages to aquifer and hydrogeologic units.

XS2D Prefix of feature classes and tables in the groundwater data model used for representing 2D cross sections.

XS2D_BoreLine Polyline feature class for representing vertical borehole data projected on a vertical plane along a section line.

XS2D_Catalog Table for managing XS2D feature classes and their association with SectionLine features. Each row in this table provides the name and role of a XS2D feature class and information about its associated SectionLine feature.

XS2D_MajorGrid and XS2D_MinorGrid Polyline feature classes representing grid lines showing the vertical and horizontal dimensions in a 2D cross section.

XS2D_Panel Polygon feature class for representing hydrogeologic units as 2D cross-section "panels." Usually a cross section will be formed by a set of XS2D_Panel features, each representing a hydrogeologic unit along a section line.

XS2D_PanelDivider Polyline feature class representing vertical lines on a cross-section plane showing the location where a section line changes direction (i.e., the location of vertices of the section line). Panel dividers are used as guides for orientation when viewing the 2D cross section.

About the authors

Gil Strassberg is the main architect of the Arc Hydro Groundwater data model. He is a senior product engineer at Aquaveo LLC, and part of the Arc Hydro Groundwater tools development team. Strassberg received his master's of science and doctorate degrees in civil engineering at the University of Texas at Austin, where he created the initial design of the groundwater data model as part of his PhD studies at the Center for Research in Water Resources. As a civil engineer, he has been involved in a variety of projects integrating GIS and water resources.

Norman L. Jones is a professor in the Department of Civil and Environmental Engineering and director of the Environmental Modeling Research Laboratory at Brigham Young University. He received his master's of science and doctorate degrees in geotechnical engineering from the University of Texas at Austin. Jones has contributed to numerous books and articles on the subject of groundwater modeling.

David R. Maidment is Hussein M. Alharthy Centennial Chair in Civil Engineering, and director of the Center for Research in Water Resources at the University of Texas at Austin. He teaches water resources engineering and conducts research on the application of GIS in water resources. Maidment edited and contributed largely to *Arc Hydro: GIS for Water Resources* (Esri Press 2002), which features the original Arc Hydro data model.

Index

A

B

C

D

E

F

G

H

I

J

K

L

M

N

O

P

R

S

T

X

Z

Related titles from Esri Press

Arc Hydro: GIS for Water Resources
ISBN: 9781589480346

This is the definitive book on the original Arc Hydro Data Model, a sophisticated template designed to help hydrologists and scientists from other disciplines develop data models in their own areas of study. This book describes how to create hydro networks of rivers and streams, define drainage areas linked via relationships to a hydro network, represent channel shape using three-dimensional models, and connect geospatial features to time-series measurements recorded at gauging sites.

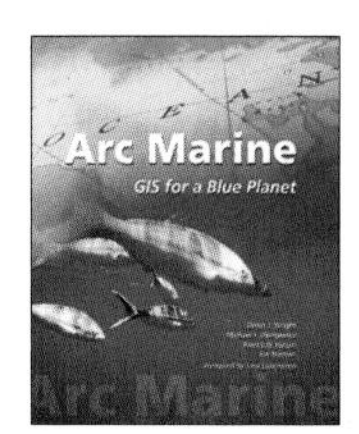

Arc Marine: GIS for a Blue Planet
ISBN: 9781589480179

Arc Marine: GIS for a Blue Planet presents the initial results of a successful effort to create and define a data model for the marine community. The data model not only provides structure to storing and analyzing marine data but helps users create maps and three-dimensional scenes of the marine environment in ways invaluable to decision making.

Smart Land-Use Analysis: The LUCIS Model
ISBN: 9781589481749

Developed by the book's authors, the LUCIS model uses the ArcGIS geoprocessing framework, particularly ModelBuilder, to analyze suitability and preference for major land-use categories, determine potential future conflict among the wcategories, and build future land-use scenarios. ArcGIS assignments are provided at various points along the way to reinforce the concepts and provide hands-on experience with LUCIS techniques.

GIS for Water Management in Europe
ISBN: 9781589480766

On the European continent, a common physical geography means common problems in natural resources and environmental management that require a unified approach to finding solutions. The case studies examined in *GIS for Water Management in Europe* recount the myriad imaginative ways that European organizations, agencies, and governments are using GIS technology to bring unity to a diverse group of problems.

Esri Press publishes books about the science, application, and technology of GIS. Ask for these titles at your local bookstore or order by calling 1-800-447-9778. You can also read book descriptions, read reviews, and shop online at www.esri.com/esripress. Outside the United States, contact your local Esri distributor.